I0560531

A History of the United States Electric Grid

By Douglas Donovan

Copyright © 2025 by Douglas Donovan

All rights reserved.

No part of this publication may be reproduced, distributed, or transmitted in any form or by any means, including photocopying, recording, or other electronic or mechanical methods, without the prior written permission of the publisher, except as permitted by U.S. copyright law. For permission requests, contact douglasdon@yahoo.com.

IBSN 979-8-9914671-7-9

First Edition 2025

Chapter 1: Introduction - The Spark of an Idea

Imagine waking up tomorrow morning to find that electricity has vanished from the world. Your alarm clock remains silent, your smartphone lies dark and lifeless, and the coffee maker sits cold and useless on your kitchen counter. As you stumble through your darkened home, reaching for light switches that no longer respond, the magnitude of our dependence on electricity becomes starkly apparent. This scenario, while impossible in our interconnected modern world, helps us appreciate the revolutionary transformation that occurred when humanity first harnessed the power of electricity.

The electrical grid that powers the United States today represents one of the most complex and remarkable engineering achievements in human history. Stretching across more than 160,000 miles of high-voltage transmission lines and connecting over 7,300 power plants to more than 140 million customers, this vast network delivers electricity with such reliability that most Americans take it for granted. Yet behind every flip of a light switch lies a fascinating story of innovation, competition, regulation, and technological evolution that spans more than a century and a half.

Understanding the Grid as a System

Before diving into its history, it's essential to understand what we mean by the "electrical grid." At its core, the grid is a sophisticated network that generates, transmits, and distributes electrical power across vast distances. Unlike other commodities, electricity cannot be easily stored in large quantities, meaning that supply must constantly match demand in real-time. This fundamental characteristic has shaped every aspect of how our electrical system has evolved.

The modern grid consists of three primary components: generation facilities that produce electricity, transmission systems that carry high-voltage power across long distances, and distribution networks that deliver electricity to homes and businesses. This system operates on the principle of alternating current (AC), a decision made over a century ago that continues to influence grid operations today. The synchronization of this entire system - ensuring that electricity flows at precisely 60 cycles per second throughout most of North America - represents a remarkable feat of engineering coordination.

The Historical Arc of Grid Development

The story of America's electrical grid unfolds in several distinct phases, each characterized by different challenges, technologies, and regulatory approaches. The earliest period, from the 1880s through the early 1900s, was marked by pioneering inventors like Thomas Edison and Nikola Tesla, whose competing visions for electrical systems would ultimately determine the technical foundation of our modern grid.

The second phase, spanning roughly from 1900 to 1935, witnessed rapid expansion and consolidation as entrepreneurs built vast utility empires. This era saw the emergence of holding companies that controlled multiple utilities across different states, leading to concerns about monopolistic practices that would ultimately trigger federal intervention.

The New Deal era of the 1930s through the 1950s brought unprecedented government involvement in the electrical industry. The creation of programs like the Tennessee Valley

Authority and the Rural Electrification Administration transformed the grid from a primarily urban phenomenon into a truly national infrastructure that reached into the most remote corners of America.

The post-World War II boom years through the 1970s represented a period of massive expansion, driven by surging demand for electricity as Americans embraced an increasingly electric lifestyle. This era saw the construction of enormous power plants, including the first generation of commercial nuclear facilities, and the development of high-voltage transmission lines that could carry power across entire regions.

The final decades of the twentieth century brought new challenges: environmental concerns, energy crises, and a growing movement toward deregulation that promised to harness market forces to improve efficiency and reduce costs. This period culminated in dramatic experiments with competitive electricity markets, some successful and others spectacularly unsuccessful, that continue to influence policy debates today.

Themes That Shape the Grid's Evolution

Throughout this historical journey, several recurring themes emerge that help explain how and why the electrical grid evolved as it did. The first is the tension between centralization and decentralization. Early electrical systems were highly localized, serving only customers within a few miles of a power plant. Over time, technological advances enabled the construction of larger, more efficient power plants that could serve broader regions, leading to increasing centralization. Today, the emergence of distributed energy resources like rooftop solar panels is once again pushing the system toward greater decentralization.

A second major theme is the ongoing debate over the proper role of government regulation versus market forces in the electrical industry. From the earliest days of electric utilities, policymakers have grappled with the challenge of ensuring reliable, affordable service while preventing monopolistic abuses. This tension has manifested in different ways throughout history, from the trust-busting efforts of the Progressive Era to the deregulation experiments of the 1990s and 2000s.

Environmental considerations represent a third crucial theme that has become increasingly important over time. What began as local concerns about air quality from coal-fired power plants has evolved into global concerns about climate change, fundamentally reshaping how we think about electricity generation and consumption.

Innovation and technological change constitute the fourth major theme. The grid's history is punctuated by breakthrough technologies - from the AC motor to nuclear power to modern computer-controlled systems - that have repeatedly transformed how electricity is generated, transmitted, and used.

The Human Impact of Electrification

Beyond the technical and economic dimensions of this story lies a profound human transformation. The arrival of electricity in American homes and communities didn't just change how people lit their houses or powered their factories; it fundamentally altered the rhythm and possibilities of daily life. Electric lighting extended the productive day, enabling new forms of work and leisure. Electric appliances transformed household labor, particularly for women, while electric transportation systems reshaped urban development patterns.

Rural electrification, in particular, represents one of the most dramatic social transformations in American history. In 1935, only about 10 percent of American farms had electricity. Within two decades, that figure had risen to over 90 percent, bringing rural Americans into the modern electrical age and helping to reduce the economic and social isolation that had long characterized rural life.

Looking Forward While Understanding the Past

As we stand at the beginning of the twenty-first century, the electrical grid faces new challenges that echo many historical themes while presenting unprecedented complexities. The integration of intermittent renewable energy sources like wind and solar power requires new approaches to grid management that would have been unimaginable to the pioneers who first strung electric wires between buildings. The emergence of electric vehicles promises to transform transportation while potentially placing new demands on an already stressed grid infrastructure.

Cybersecurity concerns highlight the grid's vulnerability to new forms of attack, while climate change raises questions about the resilience of infrastructure designed for historical weather patterns. The development of energy storage technologies offers the potential to resolve some long-standing challenges while creating new opportunities for innovation.

Understanding how we arrived at our current electrical system provides essential context for addressing these contemporary challenges. The decisions made by engineers, entrepreneurs, and policymakers over the past 140 years continue to shape the options available to us today. By examining this history thoughtfully, we can better understand not just where we've been, but where we might be heading as we work to build an electrical grid capable of meeting the needs of the twenty-first century and beyond.

This book traces that remarkable journey from Thomas Edison's first power station in lower Manhattan to the complex, interconnected system that powers modern America. It's a story of scientific breakthrough and entrepreneurial vision, of regulatory evolution and market experimentation, of triumph and occasional spectacular failure. Most importantly, it's the story of how electricity transformed from a laboratory curiosity into the invisible foundation of modern civilization.

Chapter 2: Early Innovations - Lighting the Way (1870s-1890s)

The story of America's electrical grid begins not with a single eureka moment, but with decades of scientific experimentation and practical innovation that gradually transformed electricity from a fascinating curiosity into a practical technology capable of changing the world. The foundations laid during this crucial period would determine not just the technical characteristics of our electrical system, but also the economic and regulatory structures that continue to influence the grid today.

The Scientific Foundation

Long before electricity became a commercial reality, scientists and inventors across the Atlantic world had been investigating the mysterious force that could make sparks jump between metal objects and cause compass needles to deflect. Benjamin Franklin's famous experiments with lightning in the 1740s and 1750s had established the connection between natural electrical phenomena and laboratory observations, but practical applications remained elusive for decades.

The breakthrough that made commercial electricity possible came through the work of scientists like Michael Faraday, who discovered the principle of electromagnetic induction in 1831. Faraday's insight that a moving magnet could induce an electrical current in a nearby wire provided the theoretical foundation for both electric generators and motors. However, transforming this scientific principle into practical technology required decades of additional innovation.

By the 1860s and 1870s, inventors in Europe and America were beginning to develop primitive electric generators, often called "dynamos," that could produce continuous electrical current. These early machines were crude and inefficient, but they demonstrated that electricity could be generated mechanically and used to power various devices. The challenge lay in finding applications that justified the expense and complexity of early electrical systems.

The Quest for Electric Lighting

The first commercially viable application for electricity emerged in the field of lighting. By the 1870s, inventors had developed arc lighting systems that could produce brilliant illumination by maintaining an electrical arc between two carbon electrodes. These arc lights were far brighter than any existing form of artificial lighting, but they also had significant drawbacks: they were extremely hot, produced harsh light, and required frequent maintenance to replace the carbon electrodes that were consumed in the process.

Despite these limitations, arc lighting found its first applications in situations where brightness was more important than convenience or cost. Lighthouses began installing electric arc lights that could be seen for greater distances than traditional oil lamps. Large industrial facilities and outdoor spaces like railroad yards also adopted arc lighting systems that could illuminate work areas more effectively than gas lamps.

However, arc lighting was clearly unsuitable for residential use. The harsh, flickering light and the maintenance requirements made arc lamps impractical for homes and small businesses. What was needed was a form of electric lighting that could be easily controlled, produced pleasant light, and required minimal maintenance. This need drove inventors toward the development of incandescent lighting.

Edison's Breakthrough and Business Model

Thomas Alva Edison, already famous for his invention of the phonograph and improvements to the telephone, turned his attention to electric lighting in the late 1870s. Edison understood that creating a successful incandescent lamp required solving not just the technical challenge of producing a durable filament that would glow without burning up, but also the broader challenge of creating a complete electrical system that could deliver power to multiple customers economically.

Edison's approach was systematic and comprehensive. Rather than simply inventing a light bulb, he set out to create an entire electrical supply system modeled after the existing gas lighting infrastructure in major cities. This meant developing not just improved incandescent lamps, but also generators, distribution systems, meters, switches, and all the other components needed for a complete electrical service.

After extensive experimentation with various materials for lamp filaments, Edison and his team at Menlo Park achieved a breakthrough in late 1879 with a carbonized bamboo filament that could burn for more than 1,200 hours. This represented a dramatic improvement over previous attempts and made commercial incandescent lighting economically feasible for the first time.

Equally important was Edison's choice of electrical system design. Edison opted to use direct current (DC) at relatively low voltages, typically around 110 volts. This choice was driven partly by safety considerations - lower voltages were less dangerous to users - and partly by the characteristics of early electrical equipment. Edison's DC system used a parallel circuit design that allowed individual lamps to be turned on and off independently, mimicking the functionality that customers expected from gas lighting.

The Pearl Street Station - America's First Central Power Plant

Edison's vision for electric lighting came to fruition on September 4, 1882, with the opening of the Pearl Street Station in lower Manhattan. This modest facility, housed in a renovated building at 255-257 Pearl Street, represented America's first commercial central power plant designed to serve multiple customers through an underground distribution system.

The Pearl Street Station initially served just 85 customers within a few blocks of the plant, illuminating approximately 400 incandescent lamps. The plant used six Edison "Jumbo" generators, each capable of producing about 100 kilowatts of electrical power, driven by steam engines fired by coal. Despite its small scale, the facility demonstrated the fundamental concept that would define the electrical industry for decades to come: centralized generation with distributed consumption.

Edison's choice of location was strategic. Lower Manhattan in the 1880s was already densely developed with businesses and wealthy residents who could afford the premium cost of electric lighting. The area was also served by gas lighting systems that provided a direct comparison for potential customers. By locating his power plant in the heart of America's commercial capital, Edison ensured maximum visibility for his innovation.

The initial reception was mixed but encouraging. Electric lighting was significantly more expensive than gas lighting - early customers paid about 24 cents per kilowatt-hour compared to the equivalent of about 12 cents for gas lighting. However, electric lighting offered distinct advantages: it was cleaner, safer, and more convenient than gas lighting. Electric lights didn't produce smoke or consume oxygen, they couldn't be accidentally extinguished by wind, and they could be controlled by simple switches without the need to manually light each fixture.

Technical Challenges and Solutions

The Pearl Street Station and similar early power plants faced numerous technical challenges that required innovative solutions. One of the most significant problems was voltage regulation. Edison's DC system experienced substantial voltage drops over distance, meaning that customers far from the power plant received dimmer lighting than those close by. This problem limited the practical service radius of early DC power plants to roughly half a mile.

Edison's engineers developed several solutions to address this challenge. They used large copper conductors to minimize resistance, employed a three-wire distribution system that reduced copper requirements, and eventually began installing multiple power plants in different neighborhoods to serve larger urban areas. However, these solutions were expensive and limited the economic viability of DC systems.

Another major challenge was load management. Unlike modern electrical systems, early power plants had very limited ability to store energy or adjust output quickly. This meant that generators had to be sized for peak demand, even though they operated well below capacity most of the time. Early power plant operators learned to encourage customers to use electricity during off-peak hours through differential pricing, establishing a practice that continues today.

The reliability of early electrical equipment was also problematic. Steam engines, generators, and distribution equipment all required frequent maintenance and were subject to breakdowns that could interrupt service to entire neighborhoods. Edison's company invested heavily in developing more reliable equipment and training skilled operators who could maintain the complex mechanical and electrical systems.

Expanding Beyond New York

The success of the Pearl Street Station, despite its limitations, quickly attracted attention from investors and entrepreneurs in other cities. Edison established the Edison Electric Light Company to license his technology and provide technical support for new electrical installations. By 1884, Edison companies were operating in dozens of American cities, from Boston and Philadelphia to Chicago and San Francisco.

Each new installation provided opportunities to refine the technology and business model. Edison's engineers learned to adapt their systems to local conditions, developing more efficient generators and improved distribution equipment. They also began to understand the economics of the electrical business, learning how to price service to recover the substantial capital investments required for power plants and distribution systems.

However, as the Edison system expanded, its fundamental limitations became more apparent. The short service radius of DC systems meant that large cities required multiple power plants, each serving a small area. This resulted in high capital costs and operational complexity that limited the economic attractiveness of electrical service.

Competition and Innovation

Edison's early success attracted numerous competitors who sought to improve upon his system or develop alternative approaches to electric lighting. Some focused on arc lighting systems for commercial and industrial applications, while others worked on improved incandescent technologies.

One of the most significant competitors was the Westinghouse Electric Company, founded by George Westinghouse in 1886. Westinghouse had built his reputation on innovations in railroad technology, particularly the air brake that made high-speed rail travel safer and more practical. When Westinghouse turned his attention to electricity, he brought both engineering expertise and business acumen to the challenge of improving upon Edison's approach.

Westinghouse recognized that the fundamental limitation of Edison's DC system lay in its inability to efficiently transmit power over long distances. This insight led him to investigate alternating current (AC) systems that could use transformers to step voltage up for efficient transmission and then step it back down for safe use by customers.

The competition between different electrical systems during this period was not simply a technical dispute but a fundamental disagreement about the future structure of the electrical industry. Edison's DC approach favored small, local power plants serving limited areas, which aligned with his vision of electrical utilities as local service companies similar to gas companies. Alternative approaches suggested the possibility of larger power plants serving broader regions, which would require different business models and regulatory structures.

The Foundation for Future Growth

By the end of the 1880s, electric lighting had established itself as a commercially viable technology, but it remained limited to wealthy urban customers who could afford the premium cost of electrical service. The fundamental technologies needed for widespread electrification had been demonstrated, but significant challenges remained in making electrical service more affordable and reliable.

The innovations of this foundational period established several patterns that would shape the electrical industry for decades to come. The concept of centralized generation serving multiple customers through distribution networks became the dominant model for electrical service. The need for substantial capital investment in power plants and distribution systems established the electrical business as a capital-intensive industry that required access to large amounts of financing.

The early experience with electric utilities also highlighted the importance of technical standardization and operational coordination. Electrical systems were far more complex than gas or water utilities, requiring skilled technicians and sophisticated equipment. This complexity created barriers to entry that would ultimately lead to consolidation within the industry.

Perhaps most importantly, the success of early electric lighting systems demonstrated the transformative potential of electrical technology. The convenience and versatility of electric power opened up possibilities that extended far beyond lighting, setting the stage for the broader electrification of American society that would unfold over the following decades.

As the 1890s began, electricity had moved from the laboratory to the marketplace, but its full potential remained largely unrealized. The technical and economic foundations had been established, but the great expansion of electrical service that would transform American life still lay ahead.

Chapter 3: The Dawn of Electric Streetcars and Industry

While electric lighting captured public attention and provided the first commercial application for electricity, it was the expansion of electrical technology into transportation and industry that truly began to reveal electricity's transformative potential. The 1890s witnessed a remarkable proliferation of electrical applications that extended far beyond the illumination of homes and businesses, fundamentally altering urban development patterns and industrial production methods.

The Urban Transportation Revolution

Perhaps no single application of electricity had a more dramatic impact on American cities than the electric streetcar. By the late 1880s, urban transportation was dominated by horse-drawn streetcars, which were slow, unreliable, and created significant sanitation problems in densely populated areas. The thousands of horses required to operate urban transit systems produced enormous quantities of manure that created both public health hazards and unpleasant living conditions.

The first successful electric streetcar system in the United States was developed by Frank Sprague in Richmond, Virginia, in 1888. Sprague, a former Edison employee who had left to pursue his own electrical innovations, designed a system that used overhead wires to deliver power to streetcars equipped with electric motors. This approach solved several problems that had plagued earlier attempts at electric transit: it provided a continuous source of power without the weight and complexity of onboard batteries, and it eliminated the need for dangerous third rails that posed safety hazards in urban environments.

The Richmond system was an immediate success. Electric streetcars could travel faster than horse-drawn vehicles, operate more reliably in adverse weather conditions, and provide cleaner, quieter service than steam-powered alternatives. Within just a few years, cities across America were replacing their horse-drawn transit systems with electric streetcars.

Urban Development and the Streetcar

The adoption of electric streetcars had profound implications for urban development that extended far beyond transportation. Electric streetcars made it practical for people to live greater distances from their workplaces, leading to the development of streetcar suburbs that expanded cities outward from their traditional centers.

This suburban expansion was facilitated by real estate developers who recognized the profit potential in developing residential neighborhoods connected to city centers by electric streetcar lines. Developers often invested in streetcar systems themselves, using transit connections to make previously inaccessible land suitable for residential development. This practice established a pattern of transit-oriented development that shaped American cities for decades.

The impact was particularly dramatic in rapidly growing western cities like Los Angeles, Denver, and Seattle, where electric streetcar systems were built simultaneously with residential development. In Los Angeles, Henry Huntington's Pacific Electric Railway created an extensive network of electric lines that connected communities spread across hundreds of square miles, earning the system the nickname "Big Red Cars" and establishing the framework for the region's modern urban structure.

Electric streetcars also changed the social dynamics of American cities. By making it possible for middle-class families to live in suburban neighborhoods while working in urban centers, streetcars contributed to increasing residential segregation by income. Wealthy families could afford homes in distant streetcar suburbs, while working-class families remained concentrated in neighborhoods within walking distance of employment centers.

Industrial Applications of Electric Power

While streetcars captured public attention, the adoption of electric power in industrial applications was equally significant in terms of its economic impact. Early industrial facilities had relied primarily on steam engines for power, using systems of belts and pulleys to distribute mechanical power from central steam engines to individual machines throughout a factory.

This approach had several limitations: steam engines were inefficient for variable loads, the belt and pulley systems were complex and prone to breakdowns, and the need to locate all machinery within reach of the central power source constrained factory design. Electric motors offered solutions to all of these problems.

Electric motors could be installed at each individual machine, eliminating the need for complex mechanical power distribution systems. Motors could be started and stopped instantly, allowing for more flexible production scheduling. Electric power also enabled new factory layouts that weren't constrained by the need to position all machinery within reach of a central power source.

The adoption of electric power in industry proceeded gradually through the 1890s and early 1900s. Early applications focused on specific processes where electricity offered clear advantages, such as electroplating and electric lighting within factories. As electric motors became more reliable and affordable, manufacturers began replacing steam-powered line shafts with individual electric drives for production machinery.

Mining and Electric Power

The mining industry became an early and enthusiastic adopter of electric power, particularly for underground operations where the advantages of electricity over steam power were most pronounced. Steam engines in mines required extensive ventilation systems to exhaust smoke and heat, while electric motors operated cleanly without producing emissions that could endanger miners.

Electric lighting in mines provided safer and more convenient illumination than oil lamps or candles, reducing the risk of explosions in environments where flammable gases might be present. Electric pumps could remove water from mine workings more efficiently than steam-powered alternatives, while electric hoists and conveying systems could move materials more reliably than mechanical systems.

The Comstock Lode silver mines in Nevada were among the first to adopt electric power extensively, installing electric lighting and power systems in the 1880s. Coal mines in Pennsylvania and West Virginia followed suit, using electric power to operate cutting machinery, ventilation fans, and materials handling equipment.

The Growth of Electric Appliances

As electric power became more widely available in homes and businesses, inventors began developing electrical appliances designed to take advantage of electricity's convenience and versatility. The first electric appliances were simple devices like electric irons and heating elements, but innovators quickly recognized the potential for more sophisticated applications.

Electric fans became popular in the 1890s, providing welcome relief from summer heat in homes and offices that lacked air conditioning. Electric motors were adapted to power household devices like washing machines and kitchen equipment, although these remained luxury items available only to wealthy customers due to their high cost and the limited availability of electrical service.

The development of electric appliances created new markets for electrical utilities, helping to improve the economics of power generation by increasing electricity consumption during daytime hours when lighting demand was minimal. This helped power companies make better use of their generating capacity and reduce the per-unit cost of electrical service.

Technical Innovations and Improvements

The expansion of electrical applications during the 1890s drove numerous technical innovations that improved the efficiency and reliability of electrical systems. Electric motor technology advanced rapidly as manufacturers learned to design motors optimized for specific applications, from small appliances to large industrial machinery.

Improvements in generator design increased the efficiency of power generation while reducing maintenance requirements. Better insulation materials and conductor designs enhanced the reliability of electrical distribution systems, reducing the frequency of service interruptions that had plagued early electrical installations.

The development of electrical measuring instruments allowed engineers to better understand and optimize electrical systems. Voltmeters, ammeters, and power meters enabled more precise control of electrical equipment and helped identify opportunities to improve system efficiency.

Economic and Social Implications

The expansion of electrical applications beyond lighting had profound economic implications that extended throughout American society. Electric streetcars transformed urban land values, making previously remote areas accessible for development while reducing the value of properties dependent on walking access to employment centers.

The adoption of electric power in industry improved productivity and working conditions in many applications. Electric lighting in factories enabled multiple-shift operations that increased the utilization of expensive industrial equipment. Electric motors provided more precise control over industrial processes, leading to improvements in product quality and reductions in waste.

However, the benefits of electrification were not distributed equally across American society. Electric service remained concentrated in urban areas where population density justified the substantial investment in power plants and distribution systems. Rural Americans, who comprised the majority of the population in the 1890s, had little access to electric power and continued to rely on older technologies for lighting, transportation, and industrial applications.

The capital-intensive nature of electrical systems also created barriers to competition that led to increasing consolidation within the electrical industry. The substantial investment required to build power plants and distribution systems meant that most communities could support only a single electric utility, leading to the emergence of local monopolies that would eventually require government regulation.

Setting the Stage for Further Growth

By the end of the 1890s, electricity had established itself as far more than just an improved form of lighting. The success of electric streetcars had demonstrated electricity's potential to transform transportation systems, while industrial applications showed how electric power could revolutionize manufacturing and mining operations.

These early successes created growing demand for electrical service that would drive the massive expansion of electrical infrastructure in the following decades. However, the technical limitations of existing electrical systems, particularly the short transmission distances possible with direct current technology, meant that serving this growing demand would require fundamental innovations in electrical engineering.

The stage was set for the next phase of electrical development, which would be dominated by the "War of the Currents" between competing approaches to electrical system design. The outcome of this competition would determine not just the technical characteristics of America's electrical infrastructure, but also the economic structure of the electrical industry for decades to come.

Chapter 4: The War of the Currents - AC vs. DC

The 1890s witnessed one of the most significant technical and business conflicts in American industrial history: the so-called "War of the Currents" between Thomas Edison's direct current (DC) systems and the alternating current (AC) systems promoted by George Westinghouse and Nikola Tesla. This competition was far more than a dispute between rival technologies; it was a fundamental disagreement about the future structure of the electrical industry that would determine how electricity would be generated, transmitted, and used for generations to come.

The Technical Foundation of the Conflict

To understand the significance of the AC-DC competition, it's essential to grasp the fundamental technical differences between these two approaches to electrical power. Edison's direct current systems used electrical current that flowed in one direction at constant voltage, typically around 110 volts. This approach was similar to the current produced by batteries and was compatible with the electrical equipment that Edison had developed for his lighting systems.

The primary advantage of DC systems lay in their simplicity and the maturity of the technology. Edison had spent years perfecting DC generators, motors, and distribution equipment, and his systems had proven their reliability in commercial applications. DC motors could be easily controlled and were well-suited for applications requiring variable speeds, such as electric streetcars and industrial machinery.

However, DC systems also had a critical limitation: voltage could not be easily transformed from one level to another. This meant that power had to be generated, transmitted, and used at approximately the same voltage level. Since safety considerations required relatively low voltages for customer use (around 110 volts), DC power plants had to operate at these same low voltages, which resulted in high transmission losses and limited transmission distances.

Alternating current systems, by contrast, used electrical current that periodically reversed direction, typically 60 times per second in American systems. While more complex than DC, AC had one overwhelming advantage: voltage could be easily transformed from one level to another using devices called transformers. This made it practical to generate electricity at one voltage, transmit it at a much higher voltage to minimize transmission losses, and then step it back down to safe levels for customer use.

Westinghouse and the AC Alternative

George Westinghouse came to the electrical business with a background in railroad technology and a reputation for developing practical solutions to complex engineering problems. His air brake system had revolutionized railroad safety, and he brought the same systematic approach to the challenge of improving electrical power systems.

Westinghouse recognized that the fundamental limitation of Edison's DC systems lay in their inability to transmit power economically over long distances. His investigations led him to alternating current technology, which had been developed in

Europe but had not yet been successfully commercialized in the United States.

In 1885, Westinghouse acquired the American rights to European AC patents and began developing a complete AC power system. The key breakthrough came when he hired Nikola Tesla, a brilliant Serbian immigrant who had previously worked briefly for Edison. Tesla had developed an AC motor design that solved one of the major technical challenges facing AC systems: how to use alternating current to power motors efficiently.

Tesla's polyphase motor design used multiple alternating currents that were out of phase with each other to create a rotating magnetic field that could drive motors smoothly and efficiently. This innovation made AC systems practical for industrial applications and provided a compelling alternative to DC systems for all types of electrical service.

The First AC Systems

Westinghouse's first commercial AC system was installed in Great Barrington, Massachusetts, in 1886. The system demonstrated the key advantage of AC technology: a single power plant could serve customers over a much wider area than was possible with DC systems. The Great Barrington installation used transformers to step voltage up to 3,000 volts for transmission and then step it back down to 500 volts for customer use.

The success of the Great Barrington system attracted widespread attention from electrical engineers and utility operators. Within a few years, Westinghouse was installing AC systems in cities across America, directly competing with Edison's DC systems for new electrical installations.

The competition between AC and DC systems was particularly intense in major cities where both technologies were available. In many cases, this led to the construction of competing electrical systems serving the same community, with some customers using DC power and others using AC power from different companies.

Edison's Response and the "Current War"

Thomas Edison initially dismissed AC systems as dangerous and impractical, arguing that the high voltages used in AC transmission posed unacceptable safety risks to utility workers and the public. Edison launched an aggressive campaign to discredit AC technology, highlighting accidents involving AC equipment and conducting public demonstrations designed to show the dangers of alternating current.

The most notorious aspect of Edison's anti-AC campaign involved his support for the development of the electric chair, which used AC current for executions. Edison hoped that associating AC power with death would convince the public that AC systems were too dangerous for commercial use. He even provided AC equipment to be used in the first electric chair execution in 1890.

Edison's campaign against AC technology extended beyond publicity efforts to include practical measures designed to preserve the market position of his DC systems. He developed improved DC equipment, including more efficient generators and three-wire distribution systems that reduced copper requirements. Edison also pursued patent litigation against AC system manufacturers, attempting to use intellectual property rights to limit competition.

However, Edison's efforts to preserve the dominance of DC technology were ultimately unsuccessful because they could not overcome the fundamental economic advantages of AC systems. The ability to transmit power over long distances at high efficiency gave AC systems insurmountable cost advantages for most applications.

The Chicago World's Fair - AC's Triumph

The decisive victory of AC technology came at the 1893 World's Columbian Exposition in Chicago. Westinghouse won the contract to provide electrical power for the fair, underbidding Edison's General Electric Company by nearly half. The successful operation of AC systems at the World's Fair provided a highly visible demonstration of AC technology's reliability and efficiency.

The Chicago World's Fair used more electrical power than any previous installation, with hundreds of thousands of incandescent lamps and numerous motors and other electrical devices. The fair's electrical system operated flawlessly throughout the six-month exhibition, convincing thousands of visitors and electrical professionals that AC technology was both safe and practical for large-scale applications.

The success at Chicago led directly to Westinghouse's selection to provide generators for the new hydroelectric power plant being constructed at Niagara Falls. The Niagara project represented the largest electrical installation attempted to that date and would serve as a model for future power development across America.

The Niagara Falls Project

The development of hydroelectric power at Niagara Falls provided the perfect application for AC technology's advantages. The falls offered enormous potential for power generation, but the electricity needed to be transmitted over significant distances to reach major population centers. This made the transmission efficiency of AC systems crucial to the project's economic viability.

The Niagara Falls power project was remarkable not just for its scale but also for its innovative approach to power distribution. Rather than serving local customers exclusively, the Niagara plant was designed to transmit power to Buffalo, New York, over 20 miles away - a distance that would have been impossible with DC technology of the era.

The project used 11,000-volt transmission lines, voltages far higher than had been used in previous electrical installations. This high-voltage transmission enabled efficient power delivery over long distances while demonstrating the scalability of AC systems for large regional power networks.

When the first Niagara generators began operating in 1896, they marked the beginning of a new era in electrical power. The success of the project validated the AC approach and established the technical foundation for the large-scale power systems that would develop over the following decades.

Economic and Business Implications

The victory of AC technology over DC had profound implications that extended far beyond the technical characteristics of electrical systems. AC's ability to transmit power over long distances made it economically feasible to build large, efficient power plants that could serve broad regions rather than small local areas.

This capability fundamentally changed the economics of the electrical business. Instead of requiring numerous small power plants to serve a major city, AC technology made it possible to build a few large, efficient plants that could serve the entire metropolitan area. This led to substantial economies of scale that reduced the cost of electrical service and made electricity accessible to a broader range of customers.

The regional reach of AC systems also created opportunities for utility companies to expand their service territories and achieve greater size and financial strength. This set the stage for the wave of utility consolidation that would characterize the early twentieth century, as successful companies acquired their competitors and expanded into new markets.

The End of the Current War

By the late 1890s, the outcome of the "War of the Currents" was clear: AC technology had won decisively in most applications. Even Edison's own General Electric Company, formed through a merger that reduced Edison's control over the company, began manufacturing AC equipment and abandoned the exclusive focus on DC systems that Edison had championed.

DC technology didn't disappear entirely - it remained important for specific applications like electric streetcars and industrial processes requiring precise motor control. However, AC became the standard for power generation and transmission, establishing the technical foundation that would characterize American electrical systems for the next century.

The resolution of the AC-DC competition cleared the way for rapid expansion of electrical service throughout America. With the technical approach settled, entrepreneurs and engineers could focus their efforts on building the infrastructure needed to bring electricity to communities across the nation.

Lasting Impact on Grid Development

The triumph of AC technology established several principles that continue to influence electrical grid design today. The concept of high-voltage transmission connecting large power plants to distant load centers became the fundamental organizing principle of electrical systems. The use of transformers to step voltages up and down at different points in the system enabled the development of complex, interconnected networks that could efficiently deliver power across vast distances.

Perhaps most importantly, the AC victory established the feasibility of regional and eventually national electrical networks that could share resources and improve system reliability. This capability would prove crucial as electrical demand grew and utilities sought to improve the efficiency and reliability of their systems through interconnection and coordination.

The "War of the Currents" thus represents far more than a technical dispute between competing inventors. It was a defining moment that established the fundamental architecture of America's electrical infrastructure and set the stage for the massive expansion of electrical service that would transform American society in the twentieth century.

Chapter 5: Building the First Power Plants

The victory of alternating current in the "War of the Currents" established the technical foundation for America's electrical grid, but implementing this technology on a massive scale required solving unprecedented engineering challenges. The construction of the first generation of large-scale power plants between 1900 and 1920 represents one of the most remarkable periods of technological innovation and industrial development in American history. These facilities not only had to generate electricity reliably and economically, but also had to operate within the complex social, environmental, and regulatory constraints of rapidly growing urban centers.

Engineering Challenges of Early Power Plant Design

The transition from small, local power plants serving limited areas to large regional facilities capable of serving entire metropolitan areas required fundamental innovations in power plant engineering. Early power plants like Edison's Pearl Street Station had been essentially enlarged versions of industrial steam plants, using relatively simple steam engines to drive electrical generators. However, as demand for electricity grew and AC transmission made longer-distance power delivery feasible, engineers faced the challenge of designing much larger and more sophisticated facilities.

The most fundamental challenge was achieving the reliability necessary for commercial electrical service. Unlike other industrial facilities that could shut down for maintenance or repairs without affecting external customers, power plants had to operate continuously to maintain electrical service. This requirement drove innovations in steam engine and boiler design that emphasized reliability and maintainability over pure efficiency.

Steam turbines, invented by Charles Parsons in Britain and further developed by American engineers, provided a crucial breakthrough that made large-scale power generation practical. Unlike reciprocating steam engines, which used pistons and connecting rods to convert steam pressure into rotational motion, steam turbines used high-pressure steam to spin turbine blades directly. This design was more efficient, more reliable, and capable of much higher power outputs than reciprocating engines.

The adoption of steam turbines required parallel innovations in electrical generator design. The high rotational speeds of steam turbines - often 1,800 or 3,600 revolutions per minute - required generators that could operate efficiently at these speeds while producing the 60-cycle alternating current that was becoming standard in American electrical systems. This led to the development of large, sophisticated alternators that represented significant advances in electrical engineering.

Fuel and Location Considerations

The choice of fuel and location for early power plants involved complex tradeoffs between fuel costs, transportation expenses, and proximity to customers. Coal emerged as the dominant fuel for power generation during this period, offering several advantages over alternatives. Coal was abundant in many regions of the United States, had a high energy density that made transportation economical, and could be stored easily at power plant sites to ensure fuel security.

However, coal-fired power plants required enormous quantities of fuel - a typical plant might consume several train cars of coal per day - which made transportation costs a critical factor in plant economics. This led to two distinct approaches to power plant location: some utilities built plants near coal mines and transmitted electricity to distant customers, while others located plants near population centers and transported coal to the plant sites.

Hydroelectric power offered an attractive alternative in regions with suitable water resources. Hydroelectric plants had the advantage of using a renewable fuel source that didn't require transportation, and they could operate with much smaller staffs than coal-fired plants. The success of the Niagara Falls hydroelectric project had demonstrated the feasibility of large-scale hydroelectric generation, leading to the development of similar projects across the country.

The Pacific Northwest became a center for hydroelectric development, with projects on the Columbia River and other waterways providing low-cost power that attracted energy-intensive industries like aluminum smelting. Similarly, the Tennessee Valley and other regions with suitable topography and water resources saw extensive hydroelectric development during this period.

Technical Innovations in Power Generation

The period from 1900 to 1920 witnessed remarkable innovations in power generation technology that established the foundation for modern power plants. Steam turbine technology advanced rapidly, with manufacturers like General Electric and

Westinghouse developing increasingly large and efficient units. By 1920, single steam turbines capable of generating 60,000 kilowatts were in operation - a dramatic increase from the few hundred kilowatts typical of early reciprocating steam engines.

Boiler technology also advanced significantly during this period. Early power plants had used relatively simple fire-tube boilers that were adequate for small installations but could not provide the large quantities of high-pressure steam required by modern steam turbines. The development of water-tube boilers, which circulated water through tubes exposed to hot combustion gases, enabled much higher steam pressures and temperatures while improving safety and reliability.

The integration of these technologies into complete power plant systems required sophisticated engineering and careful coordination. Power plants had to balance steam generation, turbine operation, electrical generation, and power transmission in real-time while maintaining precise control over voltage and frequency. This led to the development of complex control systems that represented early examples of industrial automation.

Cooling systems presented another significant engineering challenge. Steam turbines required large quantities of cooling water to condense steam back to liquid water for reuse in the steam cycle. This requirement led most power plants to locate near rivers, lakes, or other water sources, and it also created the first significant environmental concerns about power plant operations as heated cooling water discharge began to affect aquatic ecosystems.

Environmental and Social Impacts

The construction and operation of large power plants during this period created the first significant environmental concerns associated with electrical generation. Coal-fired power plants produced substantial quantities of smoke and ash that affected air quality in surrounding communities. The scale of these plants made their environmental impact much more significant than earlier industrial facilities.

Urban power plants faced particular challenges in managing their environmental impacts. Dense urban environments meant that power plant emissions directly affected large numbers of people, leading to complaints and eventually to municipal regulations governing power plant operations. Some utilities responded by building taller smokestacks to disperse emissions over wider areas, while others relocated new plants to less populated areas outside city centers.

Water pollution from power plant operations also became a concern as cooling water discharge affected rivers and harbors. Power plants required enormous quantities of water for cooling - a large plant might use millions of gallons per day - and the heated water discharge could raise the temperature of receiving waters enough to affect fish and other aquatic life.

The social impacts of power plant construction were equally significant. Large power plants provided substantial employment opportunities, both during construction and ongoing operations. A major power plant might employ several hundred workers directly and support additional jobs in coal mining, transportation, and related industries. This made power plant construction an important economic development tool for communities seeking to diversify their economies.

However, power plants also created social tensions, particularly when they were located near residential neighborhoods. Noise, smoke, and industrial traffic associated with power plant operations could significantly affect quality of life for nearby residents, leading to conflicts between utilities and communities that presaged later environmental justice concerns.

he Economics of Large-Scale Power Generation

The economics of early power plants were dominated by the high capital costs required for construction and the need to achieve high capacity utilization to justify these investments. A major power plant represented an investment of millions of dollars - an enormous sum at the time - which had to be recovered through electricity sales over many years.

This economic reality drove utilities to seek customers who could use large quantities of electricity consistently, helping to ensure that expensive power plant investments would be fully utilized. Industrial customers were particularly attractive because they often operated around the clock and consumed power at steady rates that matched the operating characteristics of large steam turbines.

The development of load management techniques became crucial for power plant economics. Utilities learned to encourage customers to shift electricity usage to off-peak hours through differential pricing, helping to smooth out demand variations that could leave expensive generating equipment underutilized. Time-of-use pricing and interruptible service contracts became standard tools for managing electrical loads.

The scale economies available in large power plants created strong incentives for utility consolidation and expansion. Companies that could build and operate large, efficient power plants had significant cost advantages over competitors using smaller, less efficient facilities. This economic pressure contributed to the wave of utility mergers and acquisitions that characterized the early twentieth century.

Innovations in Plant Construction and Management

The construction of large power plants required new approaches to project management and industrial construction. These facilities were among the largest and most complex industrial projects of their time, requiring coordination among architects, engineers, equipment suppliers, and construction contractors on an unprecedented scale.

Standardization became crucial for managing the complexity of power plant construction. Equipment manufacturers developed standardized designs for boilers, turbines, and generators that could be adapted to specific plant requirements while achieving economies of scale in manufacturing. This standardization also simplified plant operations and maintenance by enabling utilities to develop common procedures and training programs.

Plant management evolved from the simple approaches suitable for small facilities to sophisticated organizational structures capable of operating complex industrial facilities safely and efficiently. Large power plants required skilled operators who could manage steam systems, electrical equipment, and fuel handling systems while coordinating with system operators responsible for matching power generation to customer demand.

The development of maintenance practices suitable for large power plants became critical for achieving acceptable reliability and equipment life. Unlike smaller industrial facilities that might shut down periodically for maintenance, power plants had to maintain high availability while performing necessary maintenance on complex equipment. This led to innovations in predictive maintenance, equipment redundancy, and rapid repair techniques that became standard practice throughout the industry.

Setting the Stage for Further Expansion

By 1920, the engineering and economic principles governing large-scale power generation had been firmly established. Steam turbine technology had matured to the point where utilities could confidently build very large power plants, while improvements in transmission technology made it practical to locate these plants for optimal fuel access and environmental compatibility rather than proximity to customers.

The success of early large-scale power plants demonstrated the economic advantages of utility consolidation and regional coordination. Companies that could build and operate large, efficient power plants had clear cost advantages that translated into lower electricity prices and higher profits. This created strong market pressures that would drive the next phase of utility industry development: the rise of large utility companies and holding company empires that would dominate the electrical industry for the next several decades.

The environmental and social challenges identified during this period also established patterns that would influence utility regulation and planning for generations. The recognition that power plants created significant external effects on communities and ecosystems laid the groundwork for the environmental regulations that would become increasingly important later in the twentieth century.

Chapter 6: The Rise of Utility Companies - Consolidation and Expansion

The technical success of large-scale power generation created unprecedented opportunities for entrepreneurial vision and financial innovation in the early twentieth century. The period from 1900 to 1930 witnessed the emergence of utility empires that stretched across multiple states and served millions of customers, fundamentally transforming the electrical industry from a collection of local enterprises into a major sector of the American economy. This transformation was driven by the economics of electrical systems, but it was shaped by the ambitions of remarkable individuals who recognized the vast potential of the emerging electrical age.

The Economics Driving Consolidation

The fundamental economics of electrical systems created powerful incentives for consolidation that distinguished the utility industry from most other businesses. Unlike manufacturing companies that could achieve optimal scale with relatively modest facilities, electrical utilities found that larger systems almost invariably operated more efficiently and economically than smaller ones.

The advantages of scale in electrical systems operated at multiple levels. Large power plants could generate electricity more efficiently than small ones, reducing fuel costs per kilowatt-hour generated. Interconnected transmission systems could share generating capacity among multiple communities, reducing the total generating capacity needed to serve a given level of peak demand. Combined customer bases with different usage patterns could smooth out load variations, improving the utilization of expensive generating equipment.

These technical advantages translated into compelling economic benefits that made larger utilities more profitable and better able to provide reliable service at affordable rates. This created a competitive dynamic where larger utilities could expand by acquiring smaller competitors, offering customers better service at lower rates while earning attractive returns for investors.

The load diversity benefits of large utility systems were particularly important in the early twentieth century when electrical demand was dominated by lighting loads that varied dramatically between day and night and between summer and winter. By combining residential, commercial, and industrial customers across broad geographic areas, large utilities could achieve much more favorable load factors than small, local companies.

Samuel Insull and the Birth of the Modern Utility

No individual better exemplified the transformation of the electrical industry during this period than Samuel Insull, whose career spanned the evolution from Edison's early lighting systems to vast utility empires serving millions of customers. Insull had begun his career as Edison's personal secretary and had learned the electrical business from its founding pioneers. When he moved to Chicago in 1892 to manage the Chicago Edison Company, he brought both technical knowledge and entrepreneurial vision to one of America's most rapidly growing cities.

Insull recognized that the key to success in the electrical business lay in achieving the maximum possible utilization of expensive power plant and distribution system investments. This insight led him to pursue customers aggressively across all market segments, developing innovative rate structures and marketing programs that encouraged greater electricity usage while maintaining attractive profit margins.

Under Insull's leadership, Chicago Edison grew from a small company serving downtown Chicago to Commonwealth Edison, a major utility serving the entire Chicago metropolitan area. Insull's approach emphasized technical innovation, aggressive expansion, and sophisticated financial management that set the pattern for utility development throughout the industry.

Insull's innovations extended beyond the technical aspects of utility operations to encompass rate design, customer relations, and corporate organization. He pioneered the use of declining block rate structures that provided lower unit costs for larger customers, encouraging increased electricity consumption while maintaining overall profitability. He also developed sophisticated load research programs that enabled utilities to understand and predict customer demand patterns with unprecedented precision.

The Holding Company Model

As individual utilities grew larger and more successful, entrepreneurs recognized opportunities to achieve even greater economies of scale by combining multiple utilities under common ownership and management. This led to the development of utility holding companies that owned controlling interests in operating utilities across multiple states and regions.

The holding company model offered several advantages that made it attractive to both investors and utility managers. Holding companies could achieve economies of scale in financing by accessing capital markets more effectively than individual operating companies. They could also realize operational efficiencies by sharing engineering expertise, standardizing equipment and procedures, and coordinating system planning across multiple service territories.

Insull himself became a pioneer in holding company development, creating Middle West Utilities to own and manage utility properties throughout the Midwest and other regions. By the late 1920s, Insull's holding company empire included utilities serving customers in 32 states, representing one of the largest business enterprises in America.

The success of Insull's approach attracted numerous imitators who built their own utility holding company empires. Companies like Electric Bond and Share, American Gas and Electric, and United Corporation controlled hundreds of individual utilities serving millions of customers across the country. By 1930, a dozen major holding company groups controlled over 70 percent of America's electrical generation capacity.

Geographic Expansion and Market Development

The growth of large utility companies was facilitated by their ability to expand into new markets more effectively than smaller competitors. Large companies had access to capital, technical expertise, and management resources that enabled them to develop electrical service in communities that had been underserved by smaller local utilities.

This expansion often involved acquiring existing small utilities and upgrading their systems to modern standards. Large utilities could invest in new power plants, improved transmission and distribution systems, and better customer service that transformed the quality of electrical service available in smaller communities. These improvements often resulted in lower electricity rates despite the substantial capital investments required.

The expansion process also involved extending electrical service to previously unserved areas, particularly in rapidly growing suburban communities around major cities. Large utilities were better positioned to make the substantial investments required to build new transmission and distribution systems in areas where customer density was initially low but growth potential was high.

Rural electrification presented particular challenges and opportunities for large utilities. The low population density in rural areas made electrical service uneconomical using traditional utility business models, but some utilities recognized the potential for developing these markets through innovative approaches to system design and rate structures.

Regulatory Framework Development

The growth of large utility companies created new challenges for government regulators who had traditionally dealt with small, local utility companies that operated within single communities. Multi-state holding companies raised questions about regulatory jurisdiction and coordination that had not previously been significant concerns.

State utility commissions, which had been established in most states during the Progressive Era to regulate utility rates and service quality, found themselves dealing with companies whose operations extended far beyond state boundaries. This created opportunities for regulatory arbitrage as companies could potentially shift costs and profits among subsidiaries in different states to minimize their overall regulatory burden.

The complexity of holding company structures also made it difficult for regulators to understand the true costs and profitability of utility operations. Holding companies often provided services to their operating subsidiaries - engineering, financing, management - that made it challenging to determine whether rates charged to customers adequately reflected the actual costs of providing electrical service.

These regulatory challenges led to calls for federal oversight of utility holding companies, particularly those involved in interstate commerce. However, federal regulation of utilities was limited during this period, leaving most regulatory authority with state commissions that had limited ability to oversee multi-state operations.

Financial Innovation and Investment

The capital-intensive nature of utility development required sophisticated financial innovation that pushed the boundaries of contemporary corporate finance. Utility companies needed access to large amounts of patient capital that could be invested in long-lived assets like power plants and transmission systems that might not generate returns for many years.

Holding companies became particularly innovative in their financing approaches, using complex corporate structures that enabled them to leverage relatively small amounts of equity capital to control large utility operating companies. This financial leverage magnified returns to holding company investors during periods of growth but also created vulnerabilities that would become apparent during economic downturns.

The utility industry became one of the largest sectors of the American stock market during the 1920s as investors recognized the growth potential and stable returns available from electrical companies. Utility stocks were widely promoted as safe investments suitable for conservative investors seeking steady dividend income, attracting capital from insurance companies, pension funds, and individual investors.

The popularity of utility investments was enhanced by the industry's record of consistent growth and dividend payments. Unlike many other industries that experienced significant cyclical fluctuations, utilities showed steady increases in sales and earnings that supported regular dividend increases year after year.

Technological Standardization and Innovation

The growth of large utility systems facilitated technological standardization that improved efficiency and reduced costs throughout the industry. Large utilities had sufficient market power to influence equipment manufacturers, encouraging the development of standardized designs that could be produced in larger quantities at lower costs.

This standardization was particularly important for power generation equipment, where the development of standard steam turbine and generator designs enabled utilities to achieve greater economies of scale in procurement and maintenance. Standardization also simplified personnel training and improved system reliability by reducing the variety of equipment that operators and maintenance staff needed to understand.

Large utilities also became important centers for technological innovation, with resources to support research and development activities that were beyond the reach of smaller companies. Companies like Commonwealth Edison established engineering departments that conducted pioneering research in power system planning, load forecasting, and equipment design.

The interconnection of utility systems also drove innovations in power system protection and control that became fundamental to reliable electrical service. Large interconnected systems required sophisticated protective relaying systems that could isolate faulted equipment quickly while maintaining service to unaffected customers.

Market Power and Competition Concerns

The success of large utility companies in achieving economies of scale and providing improved service came at the cost of reduced competition in most markets. By 1930, most American communities were served by a single electrical utility that faced no direct competition for customers within its service territory.

This market structure raised concerns about the potential for monopolistic abuse, particularly as utility companies achieved greater size and financial power. Critics worried that large utilities might use their market position to charge excessive rates or provide inadequate service quality, particularly in smaller communities where their operations faced less regulatory scrutiny.

The holding company structure compounded these concerns by creating utility empires with enormous economic and political influence. The largest holding companies controlled utilities serving tens of millions of customers and represented some of the most valuable enterprises in the American economy. This concentration of economic power raised broader questions about corporate influence in American democracy.

However, defenders of large utility companies argued that the industry's natural monopoly characteristics made competition wasteful and ultimately harmful to customers. They contended that the economies of scale available in large utility systems more than compensated for the loss of competitive discipline, resulting in lower rates and better service than would be possible under competitive conditions.

Setting the Stage for Crisis

By 1930, the American utility industry had been transformed from a collection of small, local companies into a sector dominated by large holding company empires. This transformation had brought substantial benefits in the form of improved service quality, expanded availability of electrical service, and generally declining real costs of electricity.

However, the industry's structure had also created vulnerabilities that would become apparent during the economic crisis of the 1930s. The complex financial structures used by holding companies created risks that were not well understood by investors or regulators. The concentration of control in relatively few hands meant that poor decisions by holding company managers could affect millions of customers across multiple states.

The success of large utility companies had also created political tensions that would influence government policy toward the industry. Critics argued that utility empires had become too powerful and needed to be constrained through regulation or structural changes. These concerns would drive significant changes in utility regulation and industry structure during the New Deal period that followed the stock market crash of 1929.

Chapter 7: Rural Electrification - Bringing Power to the Countryside

In 1935, while American cities basked in the glow of electric lights and hummed with the productivity of electric-powered factories, nine out of ten rural homes remained shrouded in darkness, relying on kerosene lamps, hand-pumped water, and muscle-powered tools that had changed little since the nineteenth century.

This stark disparity between urban and rural America represented one of the most significant infrastructure challenges facing the nation, affecting not just the quality of life for millions of Americans but the fundamental economic development patterns that would shape the country's future.

The Rural-Urban Electric Divide

The absence of electricity in rural America was not simply a matter of technological limitation or lack of interest among farmers. Rather, it reflected the harsh economics of electrical distribution that made rural service unprofitable under the business models that had driven urban electrification. The fundamental challenge lay in the relationship between population density and the cost of electrical infrastructure.

Building electrical distribution systems required substantial investments in power lines, transformers, and other equipment that had to be amortized across the customer base served by each segment of the system. In urban areas, where customers might be separated by just a few hundred feet, the cost of serving each customer was relatively modest. In rural areas, where farms might be separated by miles of countryside, the cost per customer became prohibitively expensive.

Private utility companies, operating under the economic pressures of investor ownership, found that rural electrification projects typically could not generate sufficient revenue to justify the required infrastructure investments. A utility might need to invest thousands of dollars to connect a single rural customer who would use only modest amounts of electricity, creating an economic equation that simply didn't work under conventional utility business models.

The situation was somewhat better in certain regions of the country where geographic factors made rural electrification more feasible. In the Northwest, for example, 47.5 percent of Washington farms, 27.5 percent of Oregon farms, and 29.8 percent of Idaho farms had electricity by 1935. However, in states like Montana, only 5.5 percent of farms had electrical service, while the national average remained stubbornly low at just 10 percent.

Early Rural Electrification Efforts

Despite the economic challenges, some rural communities had begun organizing their own electrification efforts even before federal intervention. These early initiatives demonstrated both the intense desire for electrical service among rural Americans and the innovative approaches that could make rural electrification economically viable under the right circumstances.

One of the earliest examples was the Stony Run Light and Power Company, formed by farmers in Yellow Medicine County, Minnesota, in 1914. This cooperative venture initially served 26 farms and grew to 50 by 1921, purchasing power from the municipal hydroelectric plant in Granite Falls. The farmers themselves paid for all the equipment, including poles, lines, and meters, with per-farm costs ranging between $400 and $750 - a substantial investment for farming families of that era.

Similar cooperative efforts emerged in other parts of the country, often building on existing traditions of rural cooperation that had developed around grain elevators, creameries, and other agricultural enterprises. These early rural electric cooperatives proved that farmers were w lling to make significant financial sacrifices to obtain electrical service and that cooperative ownership models could make rural electrification economically feasible under certain conditions.

By 1931, approximately 63 percent of the nation's farms were paying monthly electric bills to utilities, but this figure masked significant regional variations and often represented very limited electrical service. Many rural customers had access only to basic lighting service, lacking the comprehensive electrical systems that urban customers took for granted.

The New Deal Response

President Franklin Roosevelt's administration recognized that rural electrification represented both a significant infrastructure need and an opportunity to stimulate economic recovery during the Great Depression. On May 11, 1935, Roosevelt signed Executive Order No. 7037 establishing the Rural Electrification Administration (REA) "to initiate, formulate, administer, and supervise a program of approved projects with respect to the generation, transmission, and distribution of electric energy in rural areas".

The REA was conceived as part of the broader New Deal strategy of using federal resources to address infrastructure deficits while providing employment and economic stimulus during the Depression. Congress initially allocated $100 million for the new agency - equivalent to approximately $1.88 billion in 2020 dollars - to provide loans for rural electrification projects.

The initial approach involved working with existing private utilities to extend service to rural areas using federal loan assistance. REA officials held meetings with private utility representatives, hoping to leverage existing utility expertise and infrastructure to accelerate rural electrification. However, when the utilities submitted their proposals, they exceeded the available budget and fell well short of the government's goal of widespread rural coverage.

Private utilities maintained that rural electrification would not be sustainable without additional government assistance to help finance the wiring of rural homes and the purchase of electric appliances. They argued that rural customers would not use enough electricity to make service economically viable, a position that reflected their focus on immediate profitability rather than long-term market development.

The Cooperative Solution

When private utilities proved unwilling or unable to undertake comprehensive rural electrification under the terms offered by the REA, the agency turned to an alternative approach that would ultimately prove far more successful: rural electric cooperatives. These member-owned, not-for-profit organizations combined the cooperative traditions of rural America with the federal government's financial resources to create a powerful mechanism for rural electrification.

The cooperative model offered several advantages that made it well-suited for rural electrification. Since cooperatives were owned by their members rather than outside investors, they did not need to generate profits for shareholders and could focus entirely on providing service at cost. This eliminated the profit margin that made rural service unprofitable for investor-owned utilities.

Cooperatives were also more flexible in their approach to rural electrification, often accepting lower initial usage levels and longer payback periods than private utilities would consider acceptable. Rural electric cooperative members understood that they were building infrastructure for the long term, not just pursuing immediate financial returns.

The REA provided crucial support for rural electric cooperative development beyond just financial assistance. In 1937, the agency drafted the Electric Cooperative Corporation Act, a model law that states could adopt to enable the formation and operation of not-for-profit, consumer-owned electric cooperatives. This legal framework provided the foundation for cooperative development throughout the country.

Overcoming Opposition and Challenges

The success of rural electric cooperatives did not go unopposed by private utilities, who recognized that cooperative expansion represented a threat to their potential future markets. Private companies launched various strategies to limit cooperative success, including the construction of "spite lines" that served the most lucrative rural customers while leaving less profitable areas to the cooperatives.

These spite lines were strategically designed to serve farms and businesses that could generate substantial electricity revenues while bypassing smaller customers that would be more expensive to serve. This practice deprived cooperatives of the higher-revenue customers that could help subsidize service to more remote locations, making cooperative operations more challenging financially.

Despite these obstacles, rural electric cooperatives proved remarkably successful in achieving their electrification goals. The cooperative model mobilized rural communities in ways that private utilities had never attempted, engaging local leaders and members in planning, construction, and operation of electrical systems that served their own communities.

The cooperatives also benefited from strong federal support that extended beyond financial assistance to include technical expertise, standardized equipment designs, and coordinated planning that helped individual cooperatives achieve greater efficiency and reliability. The REA became much more than a lending agency, serving as a comprehensive support system for rural electrification efforts.

The REA Circus and Public Education

Recognizing that rural electrification success required not just infrastructure construction but also education about the benefits and uses of electricity, the REA launched an innovative public education campaign known as the "REA Circus." From 1939 to 1941, this traveling exhibition crisscrossed the country promoting the advantages of electric homes and farms.

The REA Circus featured demonstrations of electric appliances, farm equipment, and home conveniences that showed rural audiences how electricity could transform their daily lives. These exhibitions were particularly important in areas where many residents had never experienced comprehensive electrical service and might be skeptical about the benefits of electrification efforts.

The educational campaign also addressed practical concerns about electrical safety, equipment installation, and efficient usage that were crucial for successful rural electrification. Many rural residents had limited experience with electrical systems and needed training to use electrical equipment safely and effectively.

Remarkable Success and Transformation

The results of the REA program and the rural electric cooperative movement exceeded even optimistic expectations. By 1956, 927 rural electric cooperatives had been created, serving 2.5 million customers. Rural electrification rates increased dramatically from 33 percent in 1940 to 96 percent in 1956. By 1953, more than 90 percent of U.S. farms had electricity, and today about 99 percent of the nation's farms have electric service.

This transformation represented one of the most successful infrastructure development programs in American history. In just two decades, rural America was brought into the electrical age through a combination of federal financial support, cooperative organization, and community commitment that overcame seemingly insurmountable economic obstacles.

The impact extended far beyond simple access to electric lighting. Rural electrification enabled the mechanization of agricultural operations through electric motors, improved food preservation through refrigeration, enhanced communication through radio and eventually television, and generally improved the quality of rural life to levels approaching urban standards.

Economic and Social Impact

Rural electrification had profound economic impacts that extended throughout American society. Electrified farms became more productive through the use of electric milking machines, feed grinders, water pumps, and other equipment that reduced labor requirements while improving output quality. This productivity improvement contributed to broader agricultural modernization that helped feed America's growing urban population.

The availability of electricity in rural areas also supported the development of rural industries and businesses that had previously been constrained by the lack of electric power. Small manufacturing operations, processing facilities, and service businesses could now operate in rural communities, providing employment opportunities that helped stem the flow of rural-to-urban migration.

Perhaps most importantly, rural electrification helped reduce the isolation that had historically characterized rural life. Electric lighting extended the productive day, while electric appliances reduced the time and effort required for household tasks. Radio and later television brought rural families into contact with national culture and information networks that had previously been largely limited to urban areas.

The Cooperative Legacy

The rural electric cooperative model established during the REA era proved remarkably durable and continues to serve rural and small-town America today. Rural electric cooperatives now serve approximately 42 million people across 56 percent of the nation's landscape, demonstrating the long-term viability of the cooperative approach to utility service.

The cooperative model also influenced utility development beyond rural electrification, providing an alternative to both private investor-owned utilities and government-owned utilities. The success of rural electric cooperatives demonstrated that consumer ownership could provide effective utility service while maintaining democratic governance and local control.

The REA program also established important precedents for federal involvement in infrastructure development that would influence later programs in telecommunications, transportation, and other sectors. The combination of federal financial assistance with local organization and ownership became a model for addressing infrastructure needs in underserved areas.

Setting the Stage for Further Development

The success of rural electrification through the REA and rural electric cooperatives resolved one of the most significant gaps in American electrical infrastructure, but it also raised broader questions about the appropriate role of government in utility development and the relative merits of different ownership models for utility service.

The demonstrated success of cooperative ownership in rural areas contributed to broader debates about public versus private ownership of utilities that would intensify during the Depression and New Deal period. If rural cooperatives could provide effective electrical service without private profit motives, some argued, perhaps similar approaches could benefit urban customers as well.

The rural electrification experience also demonstrated the importance of federal policy in addressing infrastructure needs that private markets might not serve effectively. This lesson would prove relevant to later challenges in telecommunications, broadband internet access, and other infrastructure areas where market failures might require government intervention to ensure universal access.

Chapter 8: The Debate - Private vs. Public Utilities

The remarkable success of rural electric cooperatives and the growing role of government agencies like the Tennessee Valley Authority in electrical development intensified a fundamental debate that had simmered throughout the early development of America's electrical grid: should electrical utilities be owned and operated by private companies seeking profits for their shareholders, or by public entities focused primarily on serving customers and communities? This question went far beyond technical considerations of efficiency or service quality to touch on basic philosophical differences about the role of government, the nature of essential services, and the appropriate distribution of economic power in American society.

The Philosophical Foundations

The debate over utility ownership reflected deeper ideological divisions about the proper relationship between government and private enterprise that had shaped American political discourse since the founding of the republic. Advocates of private utility ownership drew upon classical liberal economic theory, arguing that competitive markets and profit incentives provided the most effective mechanisms for ensuring efficient resource allocation and innovative service delivery.

Private ownership supporters contended that investor-owned utilities had strong incentives to operate efficiently because inefficiency would reduce profits and make it difficult to attract the capital needed for system expansion and improvement. They argued that the discipline of capital markets would ensure that utilities invested wisely and responded effectively to customer needs, since companies that failed to do so would lose market value and eventually face bankruptcy.

The private ownership model also aligned with broader American cultural values emphasizing individual initiative, limited government, and free enterprise. Supporters argued that private utilities represented the natural extension of entrepreneurial spirit that had built American industry, and that government ownership would stifle the innovation and efficiency that had made American businesses successful in other sectors.

Conversely, advocates of public ownership drew upon different philosophical traditions that emphasized the public interest and the special characteristics of essential services like electricity. They argued that electrical service was so fundamental to modern life and economic development that it should not be subject to the profit motives that might lead private companies to prioritize shareholder returns over customer welfare.

Public ownership advocates contended that utilities represented natural monopolies whose market position made traditional competitive disciplines ineffective. Since most communities could support only a single electrical distribution system, they argued, utilities would inevitably possess monopoly power that could be abused unless they were owned and controlled by the communities they served.

The Efficiency Arguments

Both sides of the ownership debate claimed that their preferred model would deliver electrical service more efficiently than the alternative, but they defined efficiency in different ways that reflected their underlying philosophical perspectives. Private ownership advocates focused primarily on economic efficiency, arguing that profit incentives would drive utilities to minimize costs and maximize productivity in ways that government-owned utilities could never achieve.

Private utilities pointed to their track record of innovation and expansion during the early decades of electrical development as evidence of their superior efficiency. They noted that private companies had developed most of the fundamental technologies that made modern electrical service possible, from Edison's lighting systems to the AC power transmission that enabled large-scale electrification.

Private ownership supporters also argued that investor-owned utilities had access to capital markets that enabled them to finance expansion and improvement projects more easily than government-owned utilities. They contended that private companies could raise capital more quickly and at lower cost than public entities that had to navigate political processes and regulatory restrictions on borrowing.

Public ownership advocates, however, challenged the private sector's definition of efficiency by pointing out that private utilities had to devote substantial resources to generating profits for shareholders rather than focusing entirely on customer service. They argued that eliminating the profit motive would enable public utilities to provide the same quality of service at lower cost by eliminating dividend payments and excessive executive compensation.

Public ownership supporters also questioned whether private utilities were actually operating as efficiently as they claimed, noting that utility holding companies had often engaged in complex financial maneuvers that seemed designed more to enrich their managers and investors than to improve customer service. The elaborate corporate structures used by holding companies raised questions about whether private ownership was actually achieving the efficiency benefits that its advocates claimed.

Service Quality and Reliability Considerations

The debate over utility ownership also encompassed disagreements about which ownership model would provide superior service quality and reliability. Private ownership advocates argued that profit incentives would drive utilities to maintain high service quality in order to satisfy customers and regulators, while competition for new service territories would encourage companies to innovate and improve their offerings.

Private utilities pointed to their record of expanding electrical service from major cities to smaller communities and eventually to rural areas as evidence of their commitment to serving customers effectively. They argued that private companies had the flexibility and resources needed to respond quickly to changing customer needs and technological opportunities.

However, public ownership advocates questioned whether private utilities were actually providing optimal service quality, particularly in rural and low-income areas where profit margins were limited. They pointed to the utilities' initial reluctance to serve rural customers as evidence that profit motives could conflict with comprehensive service provision.

The success of public power projects like the Tennessee Valley Authority provided compelling examples of how government ownership could deliver high-quality electrical service while promoting broader economic development goals. TVA had not only provided reliable electricity to a previously underserved region but had also supported flood control, navigation improvement, and economic development in ways that private utilities might not have pursued.

Rural electric cooperatives offered another model that combined public ownership principles with local control and democratic governance. The cooperatives' success in bringing electrical service to areas that private utilities had ignored demonstrated that alternative ownership models could achieve service quality goals that private companies had been unwilling or unable to meet.

Economic Development Implications

The choice between private and public utility ownership had significant implications for broader economic development that extended far beyond the electrical sector itself. Private ownership advocates argued that investor-owned utilities would attract capital investment that supported economic growth while providing employment and tax revenues that benefited local communities.

Private utilities also argued that their focus on profitability would ensure that electrical systems were developed in ways that supported the most productive economic activities. They contended that market signals reflected in electricity demand would guide utility investment toward areas and applications that generated the greatest economic value.

Public ownership supporters, however, argued that private utilities' focus on short-term profitability might conflict with longer-term economic development needs. They pointed to cases where private utilities had delayed or avoided investments in areas that could support economic growth but did not promise immediate returns.

The Tennessee Valley Authority again provided a compelling example of how public ownership could support comprehensive economic development strategies that went beyond simple profit maximization. TVA had coordinated electrical development with flood control, navigation improvement, and industrial recruitment in ways that created synergistic benefits that private utilities might not have pursued.

Chapter 9: The Great Depression and the New Deal's Response

The stock market crash of October 1929 and the Great
Depression that followed created the most severe economic crisis
in American history, fundamentally altering the relationship
between government and industry in ways that would
permanently reshape the electrical utility sector. The economic
collapse exposed the vulnerabilities of the utility holding company
empires that had dominated the industry during the 1920s, while
creating unprecedented opportunities for government
intervention in electrical development. The New Deal response to
the Depression established principles of federal involvement in
the electrical industry that continue to influence American energy
policy today.

The Collapse of Utility Empires

The Great Depression struck the electrical utility industry
with devastating force, destroying many of the holding company
empires that had seemed so successful during the prosperity of
the 1920s. The complex financial structures that had enabled
rapid expansion during good times became sources of
catastrophic vulnerability when economic conditions
deteriorated. Highly leveraged holding companies found
themselves unable to service their debts as utility revenues
declined and capital markets froze.

Samuel Insull's vast utility empire provided the most
dramatic example of how quickly and completely the industry's
giants could fall. By 1929, Insull's holding companies controlled
utilities serving customers in 32 states, representing one of the
largest business enterprises in America. However, the empire was

built on a foundation of debt and complex financial arrangements that depended on continued growth and access to capital markets.

When the stock market crashed, Insull's companies lost access to the capital they needed to refinance maturing debts and fund ongoing operations. The interconnected nature of the holding company structure meant that financial problems in one subsidiary quickly spread throughout the empire. By 1932, Insull's companies had collapsed into bankruptcy, wiping out the investments of hundreds of thousands of shareholders and revealing the extent to which utility holding companies had engaged in questionable financial practices.

The Insull collapse was not an isolated incident. Other major holding company groups, including those controlled by Associated Gas and Electric, United Corporation, and numerous smaller operators, faced similar financial difficulties that led to bankruptcies, reorganizations, and massive losses for investors. The widespread nature of these failures undermined public confidence in private utility management and created political pressure for government intervention.

Roosevelt's Response and New Deal Philosophy

Franklin D. Roosevelt came to the presidency with a fundamentally different vision of the proper relationship between government and the electrical industry than had prevailed during the 1920s. Roosevelt and his advisors believed that the Depression had revealed fundamental flaws in unregulated capitalism that required active government intervention to correct. They viewed the electrical industry as a critical infrastructure that was too important to be left entirely to private market forces.

The New Deal approach to electrical policy was shaped by Roosevelt's belief that electrical service should be treated as a public utility that served broad social and economic development goals rather than simply maximizing profits for shareholders. This philosophy led to unprecedented federal involvement in electrical development through direct government ownership and operation of electrical facilities, comprehensive regulation of private utilities, and financial support for alternative ownership models like rural electric cooperatives.

Roosevelt's personal experience with utility regulation as Governor of New York had convinced him that existing regulatory mechanisms were inadequate to control the abuses of utility holding companies. He had witnessed firsthand how complex holding company structures could be used to evade state regulation and manipulate electricity rates in ways that harmed consumers and communities.

The New Deal represented a comprehensive response to the failures of the utility industry during the 1920s and early 1930s. Rather than simply trying to repair the existing system, Roosevelt's administration sought to fundamentally restructure the relationship between government and utilities in ways that would ensure that electrical development served broader public purposes.

The Tennessee Valley Authority - A Bold Experiment

The creation of the Tennessee Valley Authority (TVA) in 1933 represented the most ambitious and controversial element of the New Deal's electrical policy. The TVA was conceived as a comprehensive regional development agency that would coordinate electrical generation with flood control, navigation improvement, and economic development across the Tennessee River basin, encompassing parts of seven states.

The TVA represented a direct challenge to private utility companies, which had long argued that government ownership of electrical facilities would be inefficient and harmful to the public interest. By demonstrating that a government agency could successfully develop and operate large-scale electrical systems while pursuing broader public purposes, the TVA became a powerful symbol of the New Deal's approach to utility policy.

The authority was granted unprecedented powers to build dams, generate electricity, and sell power to consumers throughout the Tennessee Valley region. Unlike private utilities, the TVA was not required to generate profits for shareholders and could therefore focus entirely on providing low-cost electrical service while pursuing environmental and economic development objectives that private companies might ignore.

The TVA's initial focus on dam construction served multiple purposes that illustrated the New Deal's comprehensive approach to regional development. The dams provided flood control that protected communities and agricultural areas throughout the river basin, while also creating recreational opportunities that supported tourism development. The navigation improvements facilitated commercial transportation that reduced shipping costs for regional businesses.

Most importantly for the electrical industry, the TVA's power generation facilities demonstrated that government agencies could operate electrical systems as efficiently as private companies while charging significantly lower rates. By 1945, the TVA had become one of the largest electrical utilities in the United States, serving customers throughout the Tennessee Valley at rates that were substantially below national averages.

Rural Electrification Administration - Completing the Grid

The creation of the Rural Electrification Administration (REA) in May 1935 addressed one of the most significant gaps in American electrical infrastructure: the absence of electrical service in rural areas. Roosevelt recognized that rural electrification represented both a critical infrastructure need and an opportunity to provide employment and economic stimulus during the Depression.

The REA was authorized to provide loans for rural electrification projects, with Congress initially allocating $100 million for the new agency - equivalent to approximately $1.88 billion in 2020 dollars. The agency's mission was "to initiate, formulate, administer, and supervise a program of approved projects with respect to the generation, transmission, and distribution of electric energy in rural areas".

Initial efforts to work with private utilities proved disappointing when their proposals exceeded the available budget and fell short of the government's goal of comprehensive rural coverage. Private companies maintained that rural electrification would not be economically viable without additional government assistance to finance rural home wiring and appliance purchases, reflecting their focus on immediate profitability rather than long-term market development.

The REA's eventual turn to rural electric cooperatives proved far more successful than the original approach of working with private utilities. The cooperative model aligned with New Deal principles of democratic participation and local control while providing a practical mechanism for rural communities to organize and finance their own electrification efforts.

By supporting rural electric cooperatives with low-interest loans and technical assistance, the REA achieved remarkable success in extending electrical service to previously underserved areas. The program demonstrated that alternative ownership models could succeed where private utilities had failed, providing ammunition for New Deal advocates who argued for greater government involvement in utility policy.

Public Works and Employment Programs

The New Deal's electrical initiatives were closely integrated with broader employment and public works programs designed to provide jobs for the millions of Americans who had become unemployed during the Depression. The Civilian Conservation Corps (CCC) employed hundreds of thousands of young men in reforestation and conservation projects, many of which supported electrical development by clearing rights-of-way for transmission lines and preparing sites for dam construction.

The Public Works Administration (PWA) funded numerous electrical projects, including power plant construction, transmission line development, and rural electrification initiatives. These programs provided employment for construction workers, engineers, and equipment manufacturers while building infrastructure that would support long-term economic development.

The Works Progress Administration (WPA), created in 1935, continued and expanded the public works approach by funding a wide range of infrastructure projects including electrical facilities. WPA projects built power lines, substations, and other electrical infrastructure while providing employment for skilled and unskilled workers throughout the country.

These employment programs demonstrated the New Deal's belief that government spending on infrastructure could serve multiple purposes simultaneously: providing immediate relief for unemployed workers, building facilities that would support long-term economic growth, and demonstrating the effectiveness of government planning and coordination in addressing complex social and economic challenges.

Regulatory Reform and Federal Oversight

The New Deal period also witnessed significant expansion of federal regulation over the electrical industry, reflecting Roosevelt's belief that state regulation had proven inadequate to control the abuses of utility holding companies. The Federal Power Commission, originally created in 1920 with limited authority over hydroelectric projects on federal waterways, was reorganized and granted expanded powers over interstate electrical transmission and wholesale power sales.

This expansion of federal regulatory authority represented a fundamental shift in the governance of the electrical industry. For the first time, utility companies faced comprehensive federal oversight of their interstate operations, including the authority to regulate wholesale electricity rates and transmission system access. This federal role complemented state regulation of local distribution and retail sales, creating a more comprehensive regulatory framework.

The new regulatory approach emphasized transparency and public accountability in ways that contrasted sharply with the secretive operations of many utility holding companies during the 1920s. Federal agencies were granted broad authority to investigate utility operations, examine corporate records, and require public disclosure of financial information that had previously been kept confidential.

The expansion of federal regulation also reflected the New Deal's belief that electrical service should be planned and coordinated on a regional or national basis rather than left to the uncoordinated decisions of individual companies. This philosophy would eventually lead to initiatives promoting greater interconnection among utility systems and coordinated planning of electrical development.

Economic and Social Impact of New Deal Electrical Policy

The New Deal's electrical initiatives had profound economic and social impacts that extended far beyond the immediate goal of providing employment during the Depression. The expansion of electrical service to rural areas transformed agricultural productivity and rural life in ways that supported broader economic development throughout the American economy.

Rural electrification enabled the mechanization of agricultural operations through electric motors, improved food preservation through refrigeration, and enhanced communication through radio and eventually television. These improvements increased agricultural productivity while reducing the isolation that had historically characterized rural life, helping to integrate rural communities into the broader American economy.

The TVA's regional development approach demonstrated how electrical development could be coordinated with other infrastructure investments to achieve synergistic benefits that exceeded the sum of individual projects. The authority's flood control and navigation improvements supported economic development throughout the Tennessee Valley, while low-cost electricity attracted energy-intensive industries that provided employment and economic diversification.

The success of public power projects like the TVA and rural electric cooperatives also provided compelling evidence that alternative ownership models could deliver high-quality electrical service at reasonable costs. This experience would influence utility policy debates for decades to come, providing a counterargument to private industry claims that government ownership would inevitably lead to inefficiency and poor service quality.

Long-term Consequences for Industry Structure

The New Deal's response to the Great Depression established principles and institutions that would shape the American electrical industry for generations. The expansion of federal regulatory authority created a framework for interstate coordination and planning that enabled the development of large-scale interconnected electrical systems spanning multiple states and regions.

The success of alternative ownership models like the TVA and rural electric cooperatives demonstrated that public and cooperative ownership could coexist with private utilities while serving different segments of the market effectively. This diversity of ownership models became a permanent feature of the

American electrical industry, in contrast to many other countries that adopted more uniform approaches to utility ownership.

The New Deal period also established the principle that government had a legitimate role in ensuring universal access to electrical service, even in areas where private companies found service unprofitable. This principle would later be applied to other utility services like telecommunications, establishing a precedent for government intervention to address market failures in essential service industries.

The regulatory reforms of the New Deal era created the foundation for the comprehensive federal oversight of utility holding companies that would be formalized in the Public Utility Holding Company Act of 1935. This legislation would fundamentally restructure the electrical industry by limiting the scope and complexity of utility holding companies while strengthening regulatory oversight of their operations.

Chapter 10: The Public Utility Holding Company Act (PUHCA) of 1935

The passage of the Public Utility Holding Company Act of 1935 represented one of the most dramatic and consequential pieces of legislation in the history of American business regulation. PUHCA was designed to address the fundamental structural problems that had enabled utility holding companies to engage in financial manipulation, evade regulatory oversight, and ultimately collapse during the Great Depression, wiping out the investments of millions of Americans. The act's provisions would reshape the electrical industry for the next 70 years, establishing new principles for corporate structure and regulatory oversight that extended far beyond the utility sector.

The Holding Company Problem

To understand the significance of PUHCA, it's essential to grasp the complex and often problematic structure of utility holding companies that had dominated the electrical industry during the 1920s. These organizations had evolved far beyond simple business combinations into elaborate pyramidal structures that enabled small groups of investors to control vast utility empires with relatively modest investments of their own capital.

A typical holding company structure might involve multiple layers of corporations, each owning controlling interests in the companies below it. For example, a top-level holding company might own 51 percent of a second-level holding company, which in turn owned 51 percent of a third-level holding company, which owned 51 percent of operating utilities. Through this pyramidal structure, the controlling interests of the top-level company could control operating utilities with investments representing only a small fraction of the total assets involved.

These complex structures created numerous opportunities for financial manipulation and regulatory evasion. Holding companies could shift income and expenses among their subsidiaries to minimize tax obligations and regulatory oversight. They could also engage in self-dealing transactions where holding companies charged their operating subsidiaries for services like engineering, financing, and management at inflated rates that extracted profits from regulated utilities while evading rate regulation.

The Samuel Insull empire provided the most notorious example of these abuses. By 1929, Insull's holding companies controlled operating utilities with assets worth approximately $3 billion, but this vast empire was controlled through a pyramidal structure that required relatively modest investments at the top level. The complex corporate structure made it virtually impossible for regulators or investors to understand the true financial condition of individual operating companies or the empire as a whole.

When the stock market crashed and economic conditions deteriorated, the highly leveraged structure of utility holding companies became a source of catastrophic vulnerability. Companies that appeared financially sound based on their regulated utility operations were dragged into bankruptcy by the speculative activities and excessive debt burdens of their parent holding companies.

Legislative Development and Political Opposition

The development of legislation to address utility holding company abuses became a major priority for the Roosevelt administration, which viewed these companies as symbols of the financial excesses that had contributed to the Great Depression. The legislative process was contentious and highly political, with utility holding companies mounting an intensive lobbying campaign to defeat or weaken the proposed regulations.

The initial version of what would become PUHCA was introduced in Congress in early 1935 as part of a broader package of utility regulation reforms. The legislation faced fierce opposition from utility holding companies, which argued that the proposed regulations would destroy legitimate business relationships and harm the efficiency of utility operations.

The utility industry's opposition campaign was remarkable for its scope and intensity, involving extensive lobbying, public relations efforts, and even attempts to generate fake grassroots opposition to the legislation. Holding companies organized letter-writing campaigns, sponsored advertisements, and funded speaking tours by prominent business leaders who argued that the proposed regulations would undermine American business and harm economic recovery.

Despite this opposition, public opinion strongly favored regulation of utility holding companies, whose failures during the Depression had created widespread anger among investors and consumers. The collapse of prominent holding company empires like those controlled by Insull had demonstrated the need for stronger oversight of these complex corporate structures.

The legislative debate over PUHCA also reflected broader philosophical disagreements about the proper role of government in regulating business. Supporters argued that the act was necessary to protect investors and consumers from the abuses that had characterized utility holding companies during the 1920s. Critics contended that the legislation represented excessive government interference in private business that would stifle innovation and economic growth.

Key Provisions of PUHCA

The Public Utility Holding Company Act as finally enacted contained several crucial provisions designed to address the specific problems that had been identified with utility holding company operations. The most important of these was the "death sentence" clause, which required holding companies to divest themselves of properties that were not part of integrated utility systems serving contiguous geographic areas.

This provision was designed to eliminate the sprawling, geographically dispersed utility empires that had characterized companies like those controlled by Insull. Under PUHCA, holding companies would be limited to controlling utilities that could be operated as integrated systems, eliminating the complex pyramidal structures that had enabled small groups of investors to control vast, unrelated utility properties.

The act also required holding companies to register with the Securities and Exchange Commission and submit to comprehensive federal regulation of their financial operations. This included requirements for detailed financial reporting, restrictions on transactions between holding companies and their subsidiaries, and SEC approval for major corporate actions like mergers, acquisitions, and security issuances.

PUHCA established strict limits on the corporate structures that holding companies could maintain, generally restricting them to no more than two levels between the holding company and its operating subsidiaries. This provision was designed to eliminate the complex pyramidal structures that had made it difficult for regulators and investors to understand the true financial condition of utility operations.

The act also contained provisions designed to prevent the self-dealing transactions that had been common under the previous system. Holding companies were required to charge their subsidiaries for services at cost, rather than at inflated rates that extracted excessive profits from regulated utility operations. The SEC was granted broad authority to investigate and regulate these inter-company transactions.

Implementation Challenges and Industry Resistance

The implementation of PUHCA proved to be a complex and contentious process that extended over many years as holding companies challenged the law's provisions in court while seeking to minimize their compliance obligations. Many companies argued that their existing corporate structures were already in compliance with the act's requirements, while others sought to delay implementation through legal challenges.

The Securities and Exchange Commission, which was granted primary responsibility for enforcing PUHCA, faced the enormous task of analyzing the complex corporate structures of hundreds of holding companies to determine which properties would need to be divested under the act's integration requirements. This process required detailed analysis of utility operations, financial relationships, and geographic service areas that took years to complete.

Some holding companies attempted to comply with PUHCA by restructuring their operations to meet the act's requirements while preserving as much of their existing corporate structure as possible. This often involved complex corporate reorganizations that created new holding companies serving integrated geographic areas while divesting properties that could not be justified under the act's integration standards.

Other companies chose to challenge PUHCA's constitutionality in federal court, arguing that Congress lacked the authority to regulate corporate structures as comprehensively as the act required. These legal challenges delayed implementation of the act's provisions for several years while the courts considered the constitutional questions involved.

The most significant constitutional challenge to PUHCA reached the Supreme Court in 1946 in the case of American Power & Light Co. v. SEC. The Court upheld the constitutionality of PUHCA's major provisions, ruling that Congress had the authority under the Commerce Clause to regulate utility holding companies engaged in interstate commerce. This decision cleared the way for full implementation of the act's requirements.

Industry Restructuring and Consolidation

The implementation of PUHCA led to the most comprehensive restructuring of corporate ownership in American business history, as utility holding companies were forced to divest billions of dollars worth of utility properties to comply with the act's integration requirements. This process fundamentally changed the structure of the American electrical industry while creating opportunities for new forms of industry organization.

Many of the largest utility holding companies chose to reorganize themselves into smaller, geographically integrated systems rather than fight the act's requirements. This process often involved spinning off utility properties to existing shareholders or selling them to other companies that could operate them as integrated systems.

The restructuring process also created opportunities for the formation of new utility companies that combined previously separate properties into more efficient integrated systems. Some of these combinations resulted in stronger, more efficient utilities that could provide better service at lower costs than the fragmented systems that had existed under the old holding company structure.

The elimination of complex holding company structures also had significant implications for utility financing and investment. Without the ability to leverage small equity investments across multiple levels of corporate structure, utility development required more substantial equity investments that reduced the speculative character of utility investing.

Regulatory Impact and Federal Oversight

PUHCA established the principle that federal regulation could extend beyond traditional antitrust enforcement to encompass comprehensive oversight of corporate structure and organization. The act created new regulatory tools that enabled government agencies to reshape entire industries in ways that promoted public welfare objectives.

The Securities and Exchange Commission's role in enforcing PUHCA made it one of the most powerful regulatory agencies in the federal government, with authority that extended far beyond traditional securities regulation to encompass comprehensive oversight of utility corporate structure and financial operations. This expanded regulatory role established precedents that would influence the development of federal regulation in other industries.

The success of PUHCA in eliminating the abuses associated with utility holding companies also demonstrated that comprehensive regulatory reform could be effective in addressing fundamental structural problems in major industries. This experience provided support for similar regulatory approaches in other sectors where complex corporate structures had created opportunities for abuse.

Long-term Consequences and Industry Evolution

The effects of PUHCA extended far beyond the immediate goal of eliminating utility holding company abuses to influence the fundamental character of the American electrical industry for decades. The act's integration requirements encouraged the development of larger, more efficient utility systems that could achieve greater economies of scale while serving broader geographic areas.

The elimination of speculative holding company structures also contributed to greater financial stability throughout the utility industry. Utilities operating under PUHCA's requirements were generally better capitalized and less vulnerable to the financial manipulation that had characterized the pre-1935 period.

PUHCA's success in restructuring the utility industry also influenced regulatory approaches in other sectors where similar structural problems existed. The act established principles of federal oversight that would later be applied to other regulated industries, including telecommunications, banking, and transportation.

The act remained in effect until 2005, when it was repealed as part of a broader movement toward deregulation of the electrical industry. However, many of its core principles - including requirements for transparent financial reporting and limits on conflicts of interest - were incorporated into other regulatory frameworks that continue to govern utility operations today.

Chapter 11: Post-War Expansion and Technological Advancements

The end of World War II marked the beginning of an unprecedented period of growth and technological transformation for America's electrical grid. The combination of pent-up consumer demand, rapid economic growth, technological innovations developed during the war, and massive demographic shifts created conditions that would drive the most dramatic expansion of electrical infrastructure in American history. Between 1945 and 1970, electricity consumption in the United States increased more than fourfold, requiring utilities to fundamentally rethink their approaches to power generation, transmission, and distribution.

The Post-War Economic Boom and Electricity Demand

The American economy's rapid transition from wartime production to peacetime prosperity created explosive growth in electricity demand that exceeded even the most optimistic pre-war projections. Returning veterans, armed with GI Bill benefits and steady employment in a booming economy, sought to establish households equipped with the latest electrical conveniences that had been unavailable during the Depression and war years.

The demographic phenomenon known as the baby boom contributed to massive residential construction that required extensive new electrical infrastructure. Between 1946 and 1964, approximately 76 million children were born in the United States, driving demand for new housing that reached unprecedented levels. Suburban development patterns, enabled by federal highway construction and favorable mortgage policies, created vast new residential areas that required comprehensive electrical service.

The post-war economy also witnessed rapid industrial expansion as American manufacturing capacity, enlarged during the war, was converted to peacetime production. The development of new industries like petrochemicals, aluminum production, and electronics manufacturing created large new industrial loads that demanded reliable, high-quality electrical service. These industries often required enormous amounts of electricity - an aluminum smelter might consume as much power as a medium-sized city - driving utilities to develop larger, more efficient power plants.

Commercial electricity demand also surged as the American economy shifted toward greater reliance on service industries that depended heavily on electrical equipment. Office buildings, shopping centers, and other commercial facilities required sophisticated electrical systems to support air conditioning, lighting, elevators, and the growing array of electrical office equipment that was becoming standard in American business.

Suburban Development and Electrical Infrastructure

The massive suburban development that characterized post-war America created unprecedented challenges for electrical utilities, requiring them to extend service to vast new areas while maintaining reliability standards that suburban customers expected. Unlike urban areas where electrical infrastructure could serve high customer densities, suburban development patterns required extensive distribution systems to serve relatively dispersed populations.

The Federal Housing Administration's mortgage programs and veterans' benefits made suburban homeownership accessible to millions of families who had previously been unable to afford homes. These new suburban communities were planned with electrical service as a fundamental assumption, unlike earlier residential development that had often preceded electrical availability.

Suburban homes were also designed to use far more electricity than previous residential construction. New homes typically included electric water heating, air conditioning, and a full complement of electrical appliances that made electricity consumption per household far higher than historical patterns. The "all-electric home" became a marketing concept that promoted the use of electrical energy for heating, cooling, cooking, and water heating applications that had previously relied on other fuels.

The suburban lifestyle also created new patterns of electricity use that challenged traditional utility planning. Suburban residents typically used electricity primarily during evening hours when they returned home from work, creating sharp peaks in demand that required utilities to maintain generating capacity that was used for only a few hours each day. Air conditioning adoption in suburban homes created even more extreme summer peaks that further complicated utility planning.

Technological Innovations in Power Generation

The post-war period witnessed remarkable technological advances in power generation that enabled utilities to build larger, more efficient power plants capable of serving the rapidly growing electricity demand. Steam turbine technology advanced dramatically, with manufacturers developing units capable of generating several hundred megawatts - many times larger than the largest pre-war installations.

The development of higher steam pressures and temperatures improved the efficiency of coal-fired power plants significantly, enabling utilities to generate more electricity from each ton of coal while reducing operating costs. These efficiency improvements were crucial for maintaining affordable electricity rates despite the massive investments required for system expansion.

Natural gas emerged as an important fuel for electricity generation during this period, particularly for peaking power plants that operated only during periods of high electricity demand. Gas turbines, adapted from aircraft engine technology developed during the war, could be started quickly and provided flexible generation capability that complemented large coal-fired base-load plants.

The most dramatic technological development in power generation was the commercial application of nuclear energy, which promised to provide virtually unlimited amounts of clean, affordable electricity. The first commercial nuclear power plant in the United States, Shippingport Atomic Power Station in Pennsylvania, began operation in 1957, marking the beginning of what many believed would be a nuclear-powered future for American electricity.

Nuclear power offered several advantages that made it attractive to utility planners. Nuclear plants could operate with very small fuel requirements compared to coal or oil plants, potentially reducing fuel costs and improving energy security. Nuclear technology also promised to provide large amounts of baseload generation capacity that could complement renewable sources like hydroelectric power.

High-Voltage Transmission Development

The need to connect large, efficient power plants with growing load centers often separated by hundreds of miles drove rapid development of high-voltage transmission technology. Transmission voltages increased dramatically during the post-war period, with 345-kilovolt and 500-kilovolt lines becoming common, and some experimental lines operating at 765 kilovolts.

These higher transmission voltages enabled utilities to move large amounts of power over long distances with acceptable losses, making it economical to locate power plants for optimal fuel access or environmental compatibility rather than proximity to customers. Coal-fired plants could be built near coal mines and nuclear plants could be located away from population centers while still serving distant urban areas efficiently.

The development of high-voltage transmission also enabled greater interconnection among utility systems, creating opportunities for sharing generating reserves and improving system reliability. Regional power pools emerged during this period, allowing utilities to coordinate their operations and share resources in ways that reduced costs and improved service reliability.

Extra-high-voltage transmission lines also facilitated the development of very large power plants that could serve multiple utility systems simultaneously. These large plants could achieve greater economies of scale than smaller facilities while serving broader geographic areas through the high-voltage transmission network.

The Rise of System Integration and Reliability

The growing size and complexity of electrical systems during the post-war period required new approaches to system planning and operation that emphasized reliability and coordination. The development of large interconnected systems created both opportunities for improved efficiency and new risks associated with system-wide disturbances.

The Northeast blackout of November 9, 1965, dramatically illustrated both the benefits and risks of large interconnected electrical systems. The blackout began with a relatively minor protective relay malfunction in Canada but cascaded through the interconnected system, eventually leaving much of the northeastern United States and southeastern Canada without power for up to 13 hours.

While the blackout caused significant disruption, it also demonstrated the importance of systematic approaches to reliability planning and emergency response. The incident led to the formation of regional reliability councils that developed standards and procedures for maintaining system reliability in large interconnected networks.

The post-war period also witnessed the development of sophisticated control systems that enabled utilities to manage complex electrical networks more effectively. Load dispatch centers equipped with early computer systems allowed utilities to optimize the operation of multiple power plants while monitoring system conditions in real time.

Environmental Considerations Begin to Emerge

Although environmental concerns would not become a major factor in electrical planning until the 1970s, the post-war expansion period saw the first recognition of the environmental impacts associated with large-scale electricity generation. The construction of enormous coal-fired power plants created air quality concerns in some regions, leading to early requirements for taller smokestacks and more remote plant locations.

The development of nuclear power also raised new environmental questions related to radioactive waste disposal and the potential consequences of nuclear accidents. While these concerns were initially overshadowed by enthusiasm for nuclear technology's potential, they would eventually become major factors in energy planning.

Large hydroelectric projects during this period also faced early environmental challenges, particularly related to their impacts on river ecosystems and fish migration patterns. The construction of dams on major river systems like the Columbia River created tensions between electricity generation and environmental preservation that would intensify in later decades.

Residential Electrification and Appliance Adoption

The post-war period completed the electrification of American homes that had begun during the New Deal era. By 1970, virtually all American homes had electrical service, and electricity had become the dominant energy source for many household applications that had previously relied on other fuels.

The development and mass production of electrical appliances transformed American domestic life during this period.

Automatic washing machines, clothes dryers, dishwashers, and television sets became standard equipment in middle-class homes, while air conditioning evolved from a luxury available only to the wealthy to a common feature in new home construction.

The "all-electric home" concept promoted by utilities encouraged homeowners to use electricity for space heating, water heating, and cooking applications that had traditionally used natural gas or oil. This marketing approach was supported by declining electricity costs that made electrical heating and appliances economically competitive with alternatives in many markets.

Television's rapid adoption had particularly significant implications for electrical demand patterns. By 1960, nearly 90 percent of American homes had television sets, creating new evening peaks in electricity demand as families gathered to watch popular programs. The introduction of color television in the 1960s increased the electrical requirements of television viewing significantly.

Industrial Growth and Energy-Intensive Industries

The post-war economic boom created unprecedented demand for electricity from industrial customers, particularly in energy-intensive industries like aluminum, steel, chemicals, and pulp and paper. These industries often located new facilities specifically to take advantage of low-cost electricity, creating industrial clusters around major power plants or hydroelectric facilities.

The aluminum industry became particularly important for electrical utilities because aluminum smelting requires enormous amounts of electricity - typically about 13,000 kilowatt-hours per ton of aluminum produced. Aluminum companies often negotiated special long-term contracts with utilities that provided assured markets for large blocks of electrical generation.

The chemical industry's rapid growth during the post-war period also created substantial new electrical loads, particularly for processes like chlorine production that required large amounts of electricity. The development of petrochemical industries in regions like the Gulf Coast created concentrations of electrical demand that required substantial new generating capacity.

These industrial loads were particularly valuable for utilities because they typically operated continuously, providing stable baseload demand that enabled efficient utilization of large power plants. Industrial customers also often accepted interruptible service contracts that gave utilities flexibility to reduce loads during system emergencies.

Setting the Stage for Environmental Challenges

The massive expansion of electrical generation during the post-war period established the foundation for the environmental concerns that would dominate energy policy in later decades. The construction of enormous coal-fired power plants, while successful in meeting growing electricity demand at reasonable costs, created air pollution problems that would eventually require comprehensive regulatory responses.

Nuclear power development during this period proceeded with limited attention to long-term waste disposal issues or comprehensive safety planning. The focus on rapid deployment of nuclear technology to meet growing electricity demand would eventually create legacy problems that continue to influence energy policy today.

The post-war expansion period also established patterns of electricity consumption and economic development that assumed continued growth in energy use at rates that would prove to be unsustainable when environmental and resource constraints became apparent in the 1970s. The success of this expansion created expectations about energy abundance and affordability that would be challenged by later environmental and geopolitical developments.

Chapter 12: The Environmental Awakening - New Challenges for the Grid

The 1960s and early 1970s witnessed a fundamental transformation in American environmental consciousness that would permanently alter the trajectory of electrical grid development. What had previously been viewed primarily as local nuisances - smoke from power plants, heated water discharge, and industrial pollution - began to be understood as interconnected environmental problems with far-reaching consequences for public health and ecological systems. This environmental awakening introduced new constraints and considerations into electrical planning that would challenge the industry's traditional approaches while spurring innovation in cleaner technologies.

The Emergence of Environmental Consciousness

The modern environmental movement emerged from a convergence of scientific research, public awareness campaigns, and dramatic incidents that highlighted the environmental costs of industrial development. Rachel Carson's groundbreaking book "Silent Spring," published in 1962, awakened public consciousness to the interconnected nature of ecological systems and the potential for industrial activities to cause widespread environmental damage.

Carson's work was particularly influential because it demonstrated how pollutants could move through environmental systems in ways that affected areas far removed from their original sources. This systems perspective encouraged people to think about environmental problems in regional or even global terms rather than as isolated local issues that could be addressed through traditional regulatory approaches.

The first Earth Day celebration on April 22, 1970, marked a watershed moment in American environmental politics, bringing together diverse groups of Americans who shared concerns about air and water pollution, wilderness preservation, and the long-term sustainability of industrial development patterns. An estimated 20 million Americans participated in Earth Day events, demonstrating unprecedented public support for environmental protection measures.

Television coverage of environmental problems played a crucial role in building public awareness and political support for environmental regulations. Images of polluted rivers, smoggy cities, and industrial waste sites created visceral public reactions that translated into political pressure for stronger environmental protections. The Cuyahoga River fire in Cleveland in 1969, which received extensive media coverage, became a powerful symbol of the environmental costs of unregulated industrial development.

Air Quality and Power Plant Emissions

Coal-fired power plants, which generated approximately 60 percent of American electricity by 1970, became a primary focus of environmental concern as scientific research revealed the health and environmental impacts of their emissions. These facilities were among the largest sources of air pollutants including sulfur dioxide, nitrogen oxides, particulate matter, and trace metals that contributed to both local air quality problems and regional environmental damage.

Sulfur dioxide emissions from coal plants were identified as a major contributor to acid rain, which was causing damage to forests, lakes, and buildings across broad regions of the northeastern United States and southeastern Canada. The discovery that pollutants from Midwest power plants were causing environmental damage hundreds of miles away in the Adirondack Mountains and New England demonstrated that air pollution was a regional problem requiring coordinated regulatory responses.

Urban air quality problems, including the photochemical smog that plagued cities like Los Angeles and Denver, were partly attributed to nitrogen oxide emissions from power plants that reacted with other pollutants in the presence of sunlight to form ground-level ozone. These discoveries connected power plant operations to public health problems including respiratory diseases, particularly among children and elderly adults.

The recognition of power plants as major sources of air pollution led to the development of new emission control technologies that would fundamentally alter the economics and operation of coal-fired generation. Flue gas desulfurization systems, commonly known as "scrubbers," could remove most sulfur dioxide from power plant emissions but required substantial capital investments and ongoing operating expenses that increased electricity costs.

Electrostatic precipitators and fabric filters could reduce particulate emissions dramatically but also required significant investments and careful maintenance to operate effectively. These pollution control technologies represented entirely new categories of power plant equipment that utilities had to master while maintaining reliable electricity service.

Water Quality and Thermal Pollution

Power plants' enormous requirements for cooling water created another category of environmental concerns that had not been adequately addressed during the rapid expansion of generation capacity in the post-war period. A typical large power plant might withdraw hundreds of millions of gallons of water per day for cooling purposes, with most of this water returned to rivers or lakes at temperatures significantly higher than natural conditions.

This thermal pollution could have severe impacts on aquatic ecosystems, particularly during summer months when natural water temperatures were already elevated and dissolved oxygen levels were low. Fish kills at power plant cooling water discharges became visible symbols of the environmental costs of electricity generation that attracted media attention and public concern.

The construction of nuclear power plants intensified thermal pollution concerns because nuclear plants typically discharged more waste heat per unit of electricity generated than fossil fuel plants. Nuclear plants also required enormous quantities of cooling water, with some facilities withdrawing over a billion gallons per day from nearby water bodies.

Utility responses to thermal pollution concerns included the construction of cooling towers that allowed waste heat to be dissipated to the atmosphere rather than discharged to water bodies. However, cooling towers were expensive and could create their own environmental problems including fog formation and aesthetic impacts that communities often found objectionable.

Some utilities developed innovative approaches to thermal pollution problems, such as using heated discharge water for aquaculture or greenhouse operations. However, these beneficial uses could typically utilize only a small fraction of the waste heat produced by large power plants, leaving thermal pollution as a continuing challenge for facility siting and operation.

Nuclear Power and Safety Concerns

The rapid expansion of nuclear power development during the 1960s occurred with relatively limited public attention to safety and environmental concerns, but growing environmental consciousness began to focus scrutiny on nuclear technology's potential risks and impacts. Early nuclear plants had been developed with optimistic assumptions about technology reliability and waste disposal that would prove problematic as the industry matured.

The first major nuclear accident at a commercial facility occurred at the Fermi 1 reactor in Michigan in 1966, when a piece of debris caused a partial fuel meltdown that required the reactor to be shut down for repairs. While the accident did not result in significant radiation releases, it demonstrated that serious nuclear accidents were possible and highlighted gaps in emergency planning and response capabilities.

Growing scientific understanding of radiation's health effects raised questions about the safety of routine nuclear plant operations, particularly regarding low-level radiation releases during normal operations. Environmental groups began challenging the Atomic Energy Commission's radiation exposure standards, arguing that any radiation exposure posed potential health risks that needed to be carefully weighed against nuclear power's benefits.

Nuclear waste disposal emerged as a particularly intractable problem as the accumulation of radioactive waste at reactor sites highlighted the absence of permanent disposal solutions. The assumption that waste disposal issues would be resolved as the industry developed proved to be overly optimistic, creating a legacy problem that continues to influence nuclear policy today.

The concentration of nuclear development in certain regions also raised environmental justice concerns as some communities found themselves hosting multiple nuclear facilities while others enjoyed the benefits of nuclear electricity without bearing the associated risks. These concerns would become more prominent following later nuclear accidents.

Regulatory Response and Environmental Legislation

The environmental awakening of the late 1960s and early 1970s led to the enactment of landmark environmental legislation that established new regulatory frameworks for addressing pollution problems. The National Environmental Policy Act (NEPA), signed into law on January 1, 1970, required federal agencies to consider environmental impacts in their decision-making processes and prepare environmental impact statements for major federal actions.

NEPA had particular significance for electrical development because many power projects required federal permits, financing, or other approvals that triggered environmental review requirements. The law's requirements for public participation and consideration of alternatives provided new opportunities for environmental groups and concerned citizens to influence power plant siting and design decisions.

The Clean Air Act of 1970 established national ambient air quality standards and required states to develop implementation plans for achieving these standards. Power plants, as major sources of air pollution, became subject to increasingly stringent emission limitations that required substantial investments in pollution control equipment.

The formation of the Environmental Protection Agency in 1970 consolidated federal environmental regulatory authority and provided a strong institutional advocate for environmental protection in government decision-making. The EPA's role in regulating power plant emissions would eventually extend to virtually every aspect of power plant operation, from construction through decommissioning.

The Clean Water Act of 1972 addressed water pollution problems by establishing national standards for water quality and requiring permits for industrial discharges. Power plants' cooling water systems became subject to comprehensive regulation designed to minimize their impacts on aquatic ecosystems.

Utility Industry Responses and Adaptations

The electrical utility industry initially resisted many environmental regulations, arguing that pollution control requirements would increase electricity costs and potentially compromise system reliability. However, as environmental regulations became established, utilities began developing strategies for compliance while maintaining their fundamental mission of providing reliable, affordable electricity service.

Many utilities invested heavily in research and development of cleaner technologies, working with equipment manufacturers to develop more effective and cost-efficient pollution control systems. Some companies established environmental departments staffed with specialists who could navigate the increasingly complex regulatory requirements and develop compliance strategies.

The industry also began incorporating environmental considerations into long-term planning processes, recognizing that environmental constraints would influence facility siting, technology selection, and operating procedures for decades to come. This represented a fundamental shift from earlier planning approaches that had focused primarily on meeting electricity demand at minimum cost.

Some utilities embraced environmental stewardship as a competitive advantage, promoting their investments in pollution control technology and environmental protection as evidence of their commitment to community welfare. This approach helped build public support for necessary rate increases to fund environmental compliance while positioning utilities as responsible corporate citizens.

Technological Innovation and Cleaner Generation

Environmental concerns spurred technological innovation aimed at reducing the environmental impacts of electricity generation. Research into cleaner coal combustion technologies led to the development of fluidized bed combustion systems that could burn coal more cleanly while using lower-quality fuels that were previously unsuitable for power generation.

Interest in renewable energy technologies, which had been largely dormant since the early development of hydroelectric power, began to revive as environmental concerns highlighted the advantages of generating electricity without air pollution or thermal discharge. Solar and wind technologies, while still experimental and expensive, attracted increased research and development funding.

Geothermal energy development accelerated in regions with suitable resources, offering the potential for baseload electricity generation without the environmental impacts associated with fossil fuel combustion. The success of geothermal development in California demonstrated that renewable technologies could contribute meaningfully to electricity supply under favorable conditions.

Energy efficiency and conservation technologies also attracted increased attention as environmental concerns highlighted the advantages of reducing electricity demand rather than simply building more generating capacity. Utility-sponsored programs to promote efficient appliances and building designs began to emerge as alternatives to traditional supply-side approaches to meeting growing electricity demand.

Setting the Stage for Energy Crisis

The environmental awakening of the late 1960s and early 1970s established new constraints and considerations that would influence electrical development for decades to come. However, these environmental concerns would soon be joined by energy security and economic challenges that would fundamentally alter American energy policy priorities.

The success of environmental movements in establishing new regulations and changing public attitudes toward energy development created a foundation for more comprehensive approaches to energy policy that considered environmental costs alongside economic and reliability factors. This broader perspective would prove crucial when energy crises of the 1970s forced Americans to reconsider fundamental assumptions about energy abundance and security.

The environmental awakening also established important precedents for public participation in energy planning decisions that would influence later policy debates about energy technology choices and facility siting. The recognition that energy decisions had broad environmental and social consequences that extended beyond traditional utility service territories would continue to shape energy policy development.

Chapter 13: The Energy Crisis of the 1970s - A Turning Point

The decade of the 1970s brought a series of energy crises that fundamentally altered American attitudes toward energy consumption, security, and policy. The comfortable assumptions of abundant, cheap energy that had driven post-war economic growth and electrical grid expansion were shattered by oil embargos, supply disruptions, and dramatic price increases that exposed the vulnerability of America's energy-dependent economy. These crises marked a turning point that would reshape energy policy, utility planning, and public consciousness about energy consumption for decades to come, ultimately setting the stage for the deregulation movement that would transform the electrical industry.

The First Oil Shock - 1973

The energy crisis began suddenly and dramatically on October 6, 1973, when Egypt and Syria launched a surprise attack on Israel during the Jewish holiday of Yom Kippur. The resulting Yom Kippur War triggered a chain of events that would transform global energy markets and American energy policy. On October 17, 1973, the Organization of Arab Petroleum Exporting Countries (OAPEC) announced an oil embargo against nations supporting Israel, including the United States.

The embargo had immediate and severe consequences for American energy markets. Oil prices quadrupled from approximately $3 per barrel to over $12 per barrel within a matter of months, creating economic shockwaves that extended far beyond the energy sector. Gasoline stations ran out of fuel, creating long lines of frustrated motorists and visible symbols of

America's energy vulnerability that dominated television coverage and public discourse.

While the electrical industry was less directly affected by oil supply disruptions than transportation and heating sectors - oil accounted for only about 17 percent of electricity generation in 1973 - the crisis had significant impacts on utility operations and planning. Utilities that relied heavily on oil-fired generation faced dramatic increases in fuel costs that had to be passed on to customers through rate increases, creating political and economic pressures that challenged traditional utility business models.

The crisis also disrupted utility planning assumptions that had been based on expectations of continued economic growth and stable energy prices. Many utilities found themselves with construction programs for new power plants that might no longer be economically justified given changed economic conditions and uncertain energy demand growth. More fundamentally, the oil embargo shattered American confidence in energy security and abundance that had shaped national policy since World War II.

The Natural Gas Crisis and Infrastructure Strain

While the oil embargo received the most dramatic media attention, a separate natural gas crisis developed during the mid-1970s that had even more direct impacts on electricity generation. Federal price controls on interstate natural gas sales had discouraged exploration and development of new gas supplies while encouraging consumption, creating growing shortages that became acute during the severe winter of 1976-77.

The natural gas shortage forced many industries and utilities to switch to alternative fuels, often oil or coal, increasing demand for these already expensive energy sources. Some utilities found themselves unable to obtain sufficient natural gas to operate gas-fired power plants, requiring emergency measures to maintain electrical service including the use of older, less efficient generating units.

The natural gas crisis demonstrated the interdependence of different energy sectors and the potential for shortages in one fuel to create problems throughout the energy system. It also highlighted the unintended consequences of price regulation policies that had been designed to protect consumers but had discouraged necessary investments in energy infrastructure. The severity of the 1976-77 winter created actual heating emergencies in some regions where natural gas supplies were inadequate to meet demand, leading to temporary closures of schools and businesses.

The Second Oil Shock - 1979

Just as the American economy was beginning to recover from the first oil crisis, a second energy shock struck in 1979 following the Iranian Revolution that overthrew the Shah and disrupted Iranian oil production. The loss of Iranian oil supplies, combined with panic buying and speculation, drove oil prices from approximately $15 per barrel to over $35 per barrel by 1980.

The second oil shock had more severe economic consequences than the first because it occurred against a background of already high energy prices and economic uncertainty. The crisis contributed to a deep recession in 1980-1982 that saw unemployment rise to over 10 percent, the highest levels since the Great Depression.

For the electrical industry, the second oil crisis reinforced the lessons of the first shock about the importance of fuel diversification and energy security. Utilities accelerated efforts to reduce their dependence on oil for electricity generation, investing in coal plants, nuclear facilities, and renewable energy sources that could provide more secure fuel supplies. The crisis also intensified public and political interest in energy conservation and efficiency as alternatives to simply building more generating capacity.

Government Response and Policy Revolution

The energy crises of the 1970s triggered unprecedented government intervention in energy markets as policymakers sought to address both immediate supply disruptions and longer-term energy security challenges. The creation of the Department of Energy in 1977 consolidated federal energy policy responsibilities and elevated energy issues to cabinet-level importance in government decision-making.

President Jimmy Carter declared the energy crisis "the moral equivalent of war" and called for comprehensive national efforts to reduce energy consumption and develop alternative energy sources. Carter's energy plan included proposals for increased coal use, accelerated nuclear development, conservation programs, and research into renewable energy technologies. The government's response included both regulatory measures and financial incentives designed to reshape energy consumption patterns and encourage development of domestic energy resources.

Fuel use restrictions prohibited the construction of new oil or natural gas power plants in most circumstances, encouraging utilities to rely on coal and nuclear power for new generating capacity. This policy had profound implications for the electrical industry, as it forced utilities to reconsider their generation mix and invest in technologies that might be more expensive in the short term but offered greater fuel security.

Tax incentives and research funding supported development of renewable energy technologies including solar, wind, and geothermal power. While these technologies remained expensive and limited in their applications, government support helped establish domestic industries and research capabilities that would prove important for later renewable energy development. Energy conservation programs received unprecedented government support, including building efficiency standards, appliance efficiency requirements, and funding for weatherization programs.

Utility Industry Transformation

The energy crises fundamentally altered utility industry practices and priorities in ways that would influence electrical development for decades. Traditional approaches to utility planning, which had focused primarily on meeting projected demand growth at minimum cost, were supplemented by new considerations including fuel security, supply diversification, and demand management.

Load forecasting became more sophisticated as utilities recognized that economic conditions and energy prices could significantly affect electricity demand growth. Many companies discovered that their previous assumptions about continuing rapid demand growth were no longer valid in an era of high energy prices and economic uncertainty. This realization would later contribute to the overcapacity problems that made deregulation seem attractive in the 1990s.

Fuel procurement practices were restructured to emphasize supply security and price stability rather than simply minimizing costs. Many utilities negotiated long-term coal supply contracts that provided price protection while ensuring adequate fuel supplies for their generation plants. Nuclear fuel, with its extremely low fuel costs per unit of electricity generated, became more attractive despite high capital costs and regulatory uncertainties.

Demand-side management emerged as a new utility function as companies recognized that helping customers use electricity more efficiently could be more cost-effective than building new generating capacity. Utility conservation programs, time-of-use rates, and load management systems began to be implemented as alternatives to traditional supply-side approaches. This marked the beginning of a fundamental shift in utility thinking that would eventually contribute to the development of competitive energy markets.

Nuclear Power's Complicated Promise

The energy crises initially appeared to vindicate nuclear power advocates' arguments that nuclear generation could provide energy independence and stable electricity costs. Nuclear plants' extremely low fuel costs made them increasingly attractive as oil and coal prices soared, and the technology seemed to offer a path to energy security that didn't depend on potentially unreliable foreign suppliers.

However, the nuclear industry also faced significant challenges during this period that would ultimately limit its contribution to energy independence. The Three Mile Island accident in March 1979 occurred at the height of the second oil crisis, creating a dramatic juxtaposition between energy security concerns and nuclear safety issues that complicated public and political attitudes toward nuclear power.

The accident, while it resulted in no immediate deaths or injuries, exposed serious problems with nuclear plant safety systems and emergency response procedures. The incident received intense media coverage that highlighted the potential risks of nuclear technology, creating public opposition that would make future nuclear development much more difficult and expensive. The timing of Three Mile Island was particularly unfortunate for nuclear advocates, as it occurred just as energy security concerns might have generated greater public support for nuclear expansion.

Despite these challenges, many utilities continued to pursue nuclear development during the 1980s, viewing it as essential for reducing dependence on fossil fuels. However, escalating construction costs, regulatory delays, and public opposition would eventually make nuclear power much less attractive than it had seemed during the energy crises of the 1970s.

The Birth of Energy Efficiency

One of the most significant and lasting consequences of the 1970s energy crises was the development of energy efficiency as a systematic approach to reducing energy consumption while maintaining economic productivity and quality of life. Before the crises, energy efficiency had been largely ignored because energy was cheap and abundant. The dramatic price increases of the 1970s made efficiency economically attractive for the first time.

Utilities began developing comprehensive demand-side management programs that helped customers reduce their electricity consumption through improved equipment, better building design, and more efficient operating practices. These programs represented a fundamental shift in utility business models, as companies began earning revenue by helping customers use less of their product rather than simply selling more electricity.

The federal government supported efficiency efforts through appliance efficiency standards, building energy codes, and research programs that developed new technologies and techniques for reducing energy consumption. The establishment of efficiency standards for appliances like refrigerators, air conditioners, and water heaters would eventually save consumers billions of dollars while reducing the need for new generating capacity.

Industrial energy efficiency became particularly important as high energy prices threatened the competitiveness of American manufacturing. Many industries invested heavily in more efficient equipment and processes that reduced their energy consumption per unit of output, helping maintain profitability despite higher energy costs. These efficiency improvements often had the additional benefit of reducing environmental impacts, creating synergies between energy security and environmental protection goals.

Renewable Energy's First Revival

The energy crises of the 1970s sparked the first serious interest in renewable energy technologies since the early development of hydroelectric power. Solar, wind, and other renewable technologies attracted increased research and development funding as policymakers and utilities sought alternatives to fossil fuels that could provide greater energy security.

California became a leader in renewable energy development during this period, implementing policies that encouraged solar and wind power development through tax incentives, research programs, and requirements for utility purchases of renewable electricity. The state's aggressive approach to renewable energy was driven partly by its heavy dependence on imported oil and its vulnerability to supply disruptions.

Wind power technology advanced significantly during the late 1970s and early 1980s, with the development of more efficient and reliable wind turbines that could compete economically with conventional generation in areas with good wind resources. California's wind farms in the Altamont Pass, Tehachapi Mountains, and San Gorgonio Pass became symbols of America's renewable energy potential.

Solar energy also attracted significant attention and investment during this period, with both photovoltaic and solar thermal technologies receiving government research support and private investment. While solar technologies remained expensive and limited in their applications, the foundation was laid for the rapid cost reductions and performance improvements that would make solar power competitive in later decades.

Long-term Industry Consequences

The energy crises of the 1970s had consequences that extended far beyond the immediate period of supply disruptions

114

and high prices. The crises fundamentally changed how Americans thought about energy, shifting from assumptions of abundance and security to recognition of vulnerability and the need for careful management of energy resources.

For the electrical utility industry, the crises marked the end of the era of rapid demand growth and cheap, abundant fuel supplies that had characterized the post-war period. Utilities found themselves operating in a new environment where fuel costs were volatile, demand growth was uncertain, and public policy increasingly emphasized efficiency and environmental protection rather than simply expanding supply to meet growing demand.

The industry's response to the energy crises - including investments in diverse fuel supplies, demand-side management programs, and improved efficiency - would eventually contribute to the overcapacity problems of the 1980s and 1990s that made deregulation seem attractive. The conservation programs and efficiency improvements that utilities implemented during the energy crises were so successful that electricity demand growth slowed dramatically, leaving many utilities with more generating capacity than they needed.

The crises also established important precedents for government intervention in energy markets that would influence later policy development. The recognition that energy security was a matter of national importance justified federal policies that shaped utility investment decisions, technology choices, and operating practices in ways that would have been unthinkable before the 1970s.

Perhaps most importantly, the energy crises of the 1970s began the process of reconceptualizing electricity from a simple commodity to be consumed in growing quantities to a strategic resource that needed to be managed carefully for economic, environmental, and security reasons. This new understanding would eventually contribute to the development of competitive energy markets that sought to use market mechanisms to optimize energy production and consumption in ways that traditional regulated monopolies had been unable to achieve.

Chapter 14: Deregulation Begins - The Promise of Free Energy Markets

The movement toward deregulation of the American electrical industry emerged in the 1980s and 1990s as a response to multiple converging factors: overcapacity in generation, technological advances that reduced barriers to entry, successful deregulation in other industries, and growing dissatisfaction with traditional regulated utility monopolies. The promise of deregulation was compelling - competition would drive down electricity prices, spur innovation, and provide consumers with choices that had been unavailable under the traditional regulated monopoly structure. However, the path to deregulation would prove far more complex and challenging than its advocates initially anticipated.

The Intellectual and Political Foundation

The movement toward electrical deregulation was part of a broader intellectual and political shift toward market-oriented solutions that had begun in the 1970s and accelerated during the Reagan administration. The successful deregulation of airlines, telecommunications, trucking, and other industries provided compelling examples of how competition could benefit consumers while spurring innovation and efficiency improvements.

Economic theory suggested that the electrical industry's traditional justification as a "natural monopoly" might no longer be valid for all segments of the business. While transmission and distribution systems still exhibited natural monopoly characteristics due to the inefficiency of building duplicate networks, electricity generation appeared to offer opportunities for effective competition. Advances in generating technology, particularly the development of efficient natural gas-fired combustion turbines, had reduced the minimum efficient scale for power plants while making it easier for new companies to enter the generation market.

The intellectual case for deregulation was strengthened by growing evidence that traditional cost-of-service regulation had created perverse incentives that encouraged utilities to over-invest in expensive capital projects while paying insufficient attention to operating efficiency. The "Averch-Johnson effect," identified by economists Harvey Averch and Leland Johnson, demonstrated that regulated utilities had incentives to use more capital than was economically optimal because they earned returns on their capital investments.

Academic studies of utility performance suggested that investor-owned utilities subject to traditional regulation were not necessarily more efficient than public power systems or cooperatives, challenging the assumption that private ownership automatically led to superior performance. These findings raised questions about whether the benefits of private ownership were being realized under traditional regulatory structures.

Learning from Other Industries

The success of deregulation in other network industries provided both inspiration and practical lessons for electrical industry reformers. The deregulation of telecommunications had demonstrated that it was possible to maintain universal service while introducing competition in network industries with significant infrastructure requirements. The breakup of AT&T in

1984 had created a competitive long-distance market while preserving regulated local service, providing a model that could potentially be applied to electricity.

Airline deregulation had shown that competition could lead to dramatic price reductions and service innovations, even though it also resulted in industry consolidation and the bankruptcy of some established carriers. The overall benefits of airline deregulation - lower prices, more destinations, and innovative service options - seemed to outweigh the costs of increased industry volatility and reduced service to some smaller markets.

Natural gas deregulation, implemented through the Natural Gas Policy Act of 1978 and subsequent regulatory changes, provided perhaps the most relevant precedent for electricity. The natural gas industry had successfully separated pipeline transportation from gas production and marketing, creating competitive markets for gas supply while maintaining regulated transportation services. This model suggested that it might be possible to separate electricity generation from transmission and distribution while maintaining system reliability.

However, reformers also had to grapple with the unique characteristics of electricity that made it more challenging to deregulate than other commodities. Unlike natural gas, which could be stored in pipelines and underground facilities, electricity had to be generated and consumed simultaneously, requiring constant real-time balancing of supply and demand. The physics of electrical systems also meant that problems in one part of the network could quickly spread to affect the entire system.

The National Energy Policy Act of 1992

The first major step toward electrical deregulation came with the passage of the National Energy Policy Act of 1992 (NEPA), which created the framework for competitive wholesale electricity generation markets. The act created a new class of power producers called Exempt Wholesale Generators (EWGs) that were exempt from the utility holding company restrictions of the Public Utility Holding Company Act of 1935.

EWGs could build and operate power plants without being subject to the geographic and business restrictions that applied to traditional utilities, making it easier for independent companies to enter the electricity generation market. The act also granted the Federal Energy Regulatory Commission (FERC) expanded authority to order utilities to provide transmission access to wholesale power transactions, beginning the process of separating generation from transmission.

The 1992 act represented a compromise between advocates of comprehensive deregulation and defenders of the traditional regulatory structure. By focusing on wholesale markets while leaving retail regulation to state authorities, the act enabled experimentation with competition while preserving state regulatory authority over local distribution and customer service. This approach reflected the political reality that comprehensive federal preemption of state utility regulation would face strong political opposition.

The act's provisions for EWGs were particularly important because they enabled new types of companies to enter the electricity generation market without the capital and regulatory burdens that had traditionally limited entry to large, established utilities. Independent power producers could focus exclusively on generating electricity while relying on existing utilities to provide transmission and distribution services.

FERC Order 888 - Opening the Transmission Grid

The next crucial step in electrical deregulation came in 1996 when FERC issued Order 888, which required utilities to provide "open access non-discriminatory transmission services" to all market participants. This order effectively separated transmission services from power plant ownership, disrupting the vertical integration that had characterized traditional utilities.

Order 888 was designed to address one of the fundamental barriers to competition in electricity markets: the control that incumbent utilities exercised over transmission systems that competitors needed to access customers. By requiring utilities to offer transmission services to competitors on the same terms they provided to themselves, the order aimed to create a level playing field for electricity competition.

The order also required utilities to establish Independent System Operators (ISOs) or join Regional Transmission Organizations (RTOs) that would manage transmission systems independently of generation ownership. This institutional innovation was crucial for maintaining system reliability while enabling competition, as it provided neutral entities that could balance supply and demand without favoring particular generators.

The implementation of Order 888 proved more complex than its drafters anticipated, as it required utilities to develop sophisticated systems for calculating transmission costs, managing congestion, and coordinating with multiple generation sources. Many utilities initially resisted the order's requirements, leading to lengthy legal and regulatory battles that delayed implementation.

Order 2000 and Regional Organizations

FERC's Order 2000, issued in 1999, advanced deregulation by encouraging the formation of Regional Transmission Organizations (RTOs) that would manage larger geographic areas and facilitate wholesale competition across broader regions. RTOs were envisioned as non-profit entities that would operate transmission systems independently of any market participants, enabling fair competition while maintaining reliability.

The order was motivated by recognition that transmission systems operated most efficiently when they were managed on a regional basis that transcended individual utility service territories. Regional operation could reduce costs by enabling utilities to share reserves, optimize dispatch across larger areas, and reduce the need for duplicate infrastructure. RTOs would also facilitate the development of competitive wholesale markets by providing neutral platforms for electricity trading.

However, the development of RTOs proved controversial and uneven across different regions. Some areas, particularly in the Northeast and California, moved quickly to establish RTOs that managed both transmission systems and competitive wholesale markets. Other regions, particularly in the Southeast and West, resisted RTO development due to concerns about federal oversight and the potential costs of restructuring existing arrangements.

The voluntary nature of RTO participation meant that the United States developed a patchwork of different market structures rather than the comprehensive national system that FERC had originally envisioned. This fragmented approach would create ongoing challenges for interstate electricity trade and system reliability coordination.

State-Level Deregulation Initiatives

While federal policy focused on wholesale markets, many states began implementing retail deregulation programs that would allow individual customers to choose their electricity suppliers. Pennsylvania became one of the first states to implement comprehensive deregulation in 1996, followed by California, Massachusetts, Rhode Island, and other states that sought to extend competitive benefits to retail customers.

State deregulation approaches varied significantly, reflecting different political priorities and market conditions. Some states, like Texas, implemented comprehensive restructuring that separated generation, transmission, and distribution into separate companies while establishing competitive retail markets. Others, like California, maintained integrated utilities but required them to purchase power through competitive wholesale markets while allowing some customer choice.

The diversity of state approaches created a natural experiment that would provide valuable lessons about different deregulation models. However, this diversity also created complexity for companies operating in multiple states and challenges for maintaining reliability across state boundaries where different market rules applied.

Many states that implemented deregulation did so with great optimism about the benefits that competition would provide. Proponents promised that deregulation would reduce electricity rates, improve service quality, and spur innovation in ways that traditional regulation had been unable to achieve. These promises would be severely tested by the challenges that emerged during early deregulation implementations.

Technological Enablers of Competition

The feasibility of electrical deregulation was enhanced by several technological developments that reduced barriers to entry and made competition more practical. The development of efficient natural gas-fired combustion turbines allowed new entrants to build relatively small power plants that could compete effectively with larger facilities while requiring much less capital investment than traditional coal or nuclear plants.

Advanced metering and communication technologies made it possible to track electricity consumption and transactions with the precision necessary for competitive markets. Real-time monitoring and control systems enabled system operators to manage complex networks with multiple generators and market participants while maintaining reliability standards.

Computer modeling and trading systems borrowed from financial markets enabled the development of sophisticated electricity markets that could handle the complex requirements of real-time energy balancing while providing price signals that encouraged efficient operation. These technological tools were essential for making deregulation practical, as they enabled market participants to manage risks and coordinate operations in ways that had not been possible earlier.

The development of standardized market rules and trading platforms also facilitated competition by reducing transaction costs and providing transparent price signals. Market design became a specialized field that applied economic theory and engineering principles to create trading mechanisms that could efficiently allocate electricity resources while maintaining system reliability.

Early Promise and Expectations

The initial results of deregulation efforts seemed to validate the optimism of market advocates. In Pennsylvania, one of the first states to implement comprehensive restructuring, residential customers initially experienced rate reductions of 5-10 percent while commercial and industrial customers saw even larger savings. New market entrants brought innovative service offerings and competitive pricing that had been unavailable under traditional regulation.

The wholesale markets established by RTOs and ISOs appeared to operate efficiently, with competitive pricing that reflected supply and demand conditions. Market prices during periods of abundant capacity were often lower than the regulated rates that had prevailed under traditional cost-of-service regulation, suggesting that competition was indeed delivering promised benefits.

The success of early deregulation efforts encouraged additional states to consider restructuring their electricity industries. By 2000, seventeen states plus the District of Columbia had enacted deregulation legislation or were in the process of implementing competitive markets. The momentum behind deregulation seemed unstoppable, with advocates predicting that competitive markets would soon replace regulated monopolies throughout most of the United States.

Independent power producers flourished in the new competitive environment, building new generating capacity that increased supply and put downward pressure on electricity prices. The entry of new market participants brought fresh perspectives and innovative approaches that challenged traditional utility practices while providing customers with new options and services.

Challenges and Warning Signs

Despite the initial optimism surrounding deregulation, several challenges emerged that foreshadowed the more serious problems that would develop in later years. Market design proved more complex than anticipated, with seemingly technical details having significant implications for market outcomes and participant incentives. The requirement to balance electricity supply and demand in real time created opportunities for market manipulation that had not been fully anticipated during the design process.

System reliability became more challenging to maintain in competitive markets where multiple independent generators had to be coordinated without the hierarchical control structures that had characterized integrated utilities. While competitive markets provided strong incentives for operational efficiency, they created weaker incentives for long-term reliability investments that might not generate immediate returns.

The separation of generation from transmission and distribution created new interfaces and coordination challenges that had not existed under integrated utility structures. Market participants had to develop new skills and procedures for operating in competitive environments while maintaining the high reliability standards that customers expected from electrical service.

Some market participants began to discover ways to exercise market power or manipulate market rules to increase their profits at the expense of customers or system reliability. These gaming strategies were often technically legal under existing market rules but violated the spirit of competition that deregulation was intended to promote.

Setting the Stage for Crisis

By the late 1990s, electrical deregulation had achieved significant momentum and appeared to be delivering many of the benefits that its advocates had promised. However, the complexity of electricity markets and the challenges of maintaining reliability while promoting competition were creating conditions that would soon lead to serious problems. The stage was set for the California electricity crisis that would fundamentally alter perceptions of deregulation and slow its implementation throughout the rest of the United States.

The promise of deregulation - lower prices, improved efficiency, and greater customer choice - remained compelling, but the practical challenges of implementing competitive electricity markets were proving more difficult than anticipated. The lessons learned from these early experiences would shape the next phase of industry development and influence approaches to market design for decades to come.

Chapter 15: The Rise of Independent Power Producers (IPPs)

The emergence of Independent Power Producers (IPPs) represents one of the most significant structural changes in the American electrical industry since the formation of integrated utilities in the early twentieth century. These companies, unaffiliated with traditional electric utilities, arose from the deregulation movement and technological advances that made competitive electricity generation economically viable. IPPs fundamentally challenged the vertically integrated utility model that had dominated the industry for nearly a century, introducing new business models, financing mechanisms, and competitive pressures that would reshape how electricity is generated and sold in the United States.

Defining Independent Power Producers

Independent Power Producers are companies that generate electric power for sale to utilities, end-users, or wholesale markets without being vertically integrated electric utilities themselves. Unlike traditional utilities, IPPs focus exclusively on electricity generation and do not own or operate transmission or distribution systems. This specialization allows IPPs to concentrate their expertise and capital on power generation while relying on other entities for transmission and customer service functions.

The National Energy Policy Act of 1992 created the legal framework that enabled IPPs to flourish by establishing a new category of power producers called Exempt Wholesale Generators (EWGs). EWGs were exempt from the restrictions of the Public Utility Holding Company Act of 1935, which had limited the geographic scope and business activities of utility companies. This exemption allowed IPPs to build and operate power plants without the regulatory constraints that had historically limited entry into the electricity generation business.

IPPs differ from traditional utilities in several fundamental ways. While regulated utilities are guaranteed cost recovery and a reasonable return on investment through the rate-making process, IPPs must compete for customers and assume full market risk for their investments. This market exposure creates strong incentives for efficiency and cost control but also exposes IPPs to greater financial risk than regulated utilities face.

The business models employed by IPPs vary significantly depending on their target markets and strategies. Some IPPs focus on developing and operating large baseload power plants that sell electricity through long-term contracts with utilities. Others specialize in peaking plants that operate only during periods of high demand, capturing premium prices during peak hours. Still others develop renewable energy projects that benefit from government incentives and environmental regulations.

Technological Drivers of IPP Development

The rise of IPPs was enabled by significant technological advances that reduced barriers to entry in electricity generation. The development of efficient natural gas-fired combustion turbines was particularly important, as these units could be built relatively quickly and at much lower capital cost than traditional coal or nuclear plants. Combined-cycle plants, which use both gas turbines and steam turbines to achieve higher efficiency, allowed IPPs to compete effectively with existing utility plants while requiring much less capital investment.

These technological advances reduced the minimum efficient scale for power plants, making it possible for smaller companies to compete with large integrated utilities. A new IPP could build a 500-megawatt combined-cycle plant for a fraction of the capital required for a large coal or nuclear facility, while achieving comparable or superior efficiency and environmental performance. This democratization of power generation technology was crucial for enabling competition in the electricity industry.

Improvements in plant automation and remote monitoring capabilities also favored IPPs by reducing the personnel requirements for power plant operation. Modern power plants could be operated with much smaller staffs than earlier facilities, reducing ongoing operating costs and making it easier for independent companies to manage multiple facilities efficiently. Standardized designs for gas turbines and other equipment further reduced costs and risks for IPP developers.

The development of sophisticated financial instruments and risk management tools enabled IPPs to manage the financial risks associated with competitive electricity generation. Power purchase agreements, fuel supply contracts, and financial derivatives allowed IPPs to hedge various risks while providing certainty to investors and lenders. These financial innovations were essential for making IPP development attractive to the capital markets.

Early IPP Development and Market Entry

The first generation of IPPs emerged in the 1980s and early 1990s, primarily developing projects that sold power to utilities through long-term contracts. Many of these early IPPs were subsidiaries of large corporations with experience in energy or construction industries, such as Bechtel, General Electric, and various oil companies. These companies brought engineering expertise, financial resources, and project management skills that were essential for successful power plant development.

Qualifying Facilities (QFs) under the Public Utility Regulatory Policies Act (PURPA) of 1978 provided an early market opportunity for independent power development. QFs, which included small power producers and cogeneration facilities, were entitled to sell power to utilities at avoided cost rates. While this market was limited, it provided valuable experience for companies that would later become major IPPs in deregulated markets.

California's Standard Offer contracts in the 1980s created one of the first large-scale markets for independent power. These contracts offered long-term, fixed-price agreements that enabled IPPs to obtain financing for new power plants while providing utilities with predetermined capacity additions. The program was highly successful in attracting new generation capacity, though some contracts proved problematic when natural gas prices declined and the fixed payments became above-market.

The success of early IPP projects demonstrated that independent companies could develop and operate power plants as efficiently as traditional utilities. These projects often used innovative financing structures, including non-recourse project finance that limited investor risk to the specific project rather than requiring corporate guarantees. This financing approach would become standard practice for IPP development and would influence utility financing practices as well.

Competitive Strategies and Business Models

As competitive markets developed in the 1990s, IPPs adopted various strategies to differentiate themselves and compete effectively with incumbent utilities. Many IPPs focused on operational excellence, using advanced technology and efficient management practices to achieve lower operating costs than traditional utilities. The ability to make decisions quickly without regulatory approval gave IPPs significant advantages in responding to market opportunities and changing conditions.

Merchant power plants, which sell electricity into competitive wholesale markets without long-term contracts, represented a new and riskier business model that some IPPs embraced. These facilities bet on their ability to generate profits through superior market timing and operational flexibility, rather than relying on the security of long-term contracts. While merchant plants faced greater financial risks, they also had the potential for higher returns if they could operate efficiently in volatile markets.

Some IPPs specialized in particular technologies or market segments where they could develop competitive advantages. Renewable energy IPPs focused on wind, solar, and other clean technologies that benefited from environmental regulations and government incentives. Peaking power specialists developed plants designed to operate only during periods of high demand and high prices, maximizing revenue per hour of operation while minimizing capital requirements.

Geographic diversification became another important strategy for larger IPPs, as it allowed them to spread risks across multiple markets while taking advantage of regional differences in supply, demand, and regulation. Companies like Calpine and NRG Energy built portfolios of plants in multiple states and regions, reducing their exposure to problems in any single market while gaining expertise in different regulatory environments.

Impact on Electricity Markets

The entry of IPPs into electricity markets had profound effects that extended far beyond simply adding new generating capacity. Competition from IPPs put pressure on existing utilities to improve their efficiency and reduce costs, as utility plants that had been economic under regulated monopoly conditions suddenly faced competition from more efficient alternatives.

IPPs brought new technologies and operating practices that often outperformed existing utility plants. The higher efficiency of combined-cycle plants operated by IPPs forced utilities to reconsider the economics of their older, less efficient facilities. Some utilities found that purchasing power from IPPs was more cost-effective than continuing to operate their own plants, leading to plant retirements and capacity sales.

The flexibility and market responsiveness of IPPs also influenced electricity pricing patterns in competitive markets. IPPs were often quicker than utilities to adjust their operations in response to changing fuel prices, environmental regulations, or market conditions. This responsiveness helped make wholesale electricity markets more efficient while providing price signals that encouraged appropriate investment and consumption decisions.

Competition from IPPs also spurred innovation in power plant technology and operations. IPPs had strong incentives to adopt new technologies that could provide competitive advantages, while utilities operating under traditional regulation had weaker incentives to take risks on unproven technologies. This dynamic helped accelerate the development and deployment of cleaner and more efficient generation technologies.

Financial Innovation and Capital Markets

The IPP industry pioneered new approaches to power plant financing that differed significantly from traditional utility finance. Project finance, where loans are secured by the cash flows from specific projects rather than corporate balance sheets, became the dominant financing method for IPP development. This approach allowed companies to develop multiple projects while limiting their financial exposure to any single facility.

The development of sophisticated risk management tools enabled IPPs to operate in competitive markets while providing investors with acceptable risk profiles. Power purchase agreements, fuel supply contracts, and financial derivatives allowed IPPs to hedge various risks while maintaining the flexibility needed to compete effectively. These financial innovations influenced utility industry practices and contributed to the development of electricity derivatives markets.

Capital markets began to view IPPs as a distinct asset class with characteristics different from traditional utilities. While IPPs faced greater market risk than regulated utilities, they also offered the potential for higher returns and growth that attracted different types of investors. The success of IPP financing helped demonstrate that competitive electricity generation could attract private capital without regulatory guarantees.

The IPP industry also developed new metrics and analytical frameworks for evaluating power generation investments. Heat rates, capacity factors, availability rates, and other performance measures became standard benchmarks for comparing different projects and technologies. These metrics provided greater transparency about power plant performance than had been available under traditional utility regulation.

Challenges and Market Risks

Despite their successes, IPPs faced significant challenges and risks that traditional utilities did not encounter. Market risk was perhaps the most significant challenge, as IPPs had to compete for customers and were exposed to volatile electricity prices without the regulatory protections that utilities enjoyed. Merchant power plants were particularly vulnerable to market downturns that could make their operations unprofitable.

The cyclical nature of power markets created boom and bust periods that severely tested IPP business models. During periods of tight supply and high prices, IPPs could earn substantial returns that attracted new investment and development. However, the resulting capacity additions often led to oversupply and low prices that made many projects unprofitable. The power industry's capital intensity and long asset lives made these cycles particularly damaging.

Regulatory uncertainty posed another significant challenge for IPPs, as changes in environmental regulations, market rules, or government policies could dramatically affect project economics. The long development and operating periods for power plants meant that IPPs had to make investment decisions based on projections of regulatory conditions that might change significantly over time.

Construction and operating risks that had been shared with ratepayers under traditional utility regulation became the sole responsibility of IPPs in competitive markets. Cost overruns, performance shortfalls, or technical problems that might have been recoverable through rate increases for regulated utilities could bankrupt independent developers. This risk allocation created strong incentives for efficient project development but also led to some spectacular failures.

Case Studies of Success and Failure

The IPP industry includes examples of both remarkable successes and dramatic failures that illustrate the opportunities and risks of competitive power generation. Calpine Corporation, founded in 1984, became one of the largest IPPs in the United States by focusing on efficient natural gas-fired plants and geographic diversification. The company's success demonstrated that independent companies could compete effectively with traditional utilities while earning attractive returns for investors.

However, Calpine also illustrates the risks facing IPPs, as the company filed for bankruptcy in 2005 due to high debt levels and unfavorable market conditions. The bankruptcy highlighted how quickly market conditions could change and how leverage could amplify both gains and losses in competitive electricity markets. Calpine eventually emerged from bankruptcy and continued operations, but the experience demonstrated the cyclical nature of IPP profitability.

AES Corporation represents another significant IPP success story, growing from a small consulting firm in 1981 to become one of the world's largest independent power companies. AES pioneered international power development and innovative financing techniques while maintaining a focus on operational excellence and environmental stewardship. The company's growth demonstrated the global opportunities available to successful IPPs.

Enron's power trading and generation business provides a cautionary tale about the risks of aggressive strategies and financial manipulation in competitive power markets. While Enron was not exclusively an IPP, its power business employed many IPP strategies and innovations. The company's collapse demonstrated how quickly success could turn to failure when business models were based on unsustainable financial engineering rather than operational excellence.

Long-term Impact on Industry Structure

The rise of IPPs fundamentally altered the structure of the American electrical industry by breaking the monopoly that integrated utilities had maintained over power generation for nearly a century. By 2000, IPPs owned approximately 15 percent of U.S. generating capacity, with their share continuing to grow in subsequent years as deregulation expanded and new projects were developed.

The success of IPPs validated the economic theory that electricity generation could be competitive while maintaining system reliability. This validation provided crucial support for continued deregulation efforts while demonstrating that the natural monopoly rationale for utility integration did not extend to all aspects of the electricity business.

IPPs also changed the competitive dynamics of electricity markets by introducing new business models, technologies, and operating practices that incumbent utilities had to match or exceed. This competition drove improvements in efficiency and innovation throughout the industry while providing consumers with potential benefits from lower costs and improved service.

The IPP model also influenced international electricity development, as American companies exported their expertise and business models to other countries undergoing power sector reform. The success of American IPPs helped establish competitive electricity generation as a viable alternative to state-owned or regulated monopoly utilities in markets around the world.

The rise of IPPs marked a fundamental transformation in the American electrical industry that would continue to influence industry development for decades to come. While competitive generation would face challenges and setbacks, the IPP model had permanently altered the landscape of American electricity markets and established competition as a viable alternative to traditional utility monopolies.

Chapter 16: California's Deregulation Experiment - A Cautionary Tale

California's ambitious attempt to deregulate its electricity markets in the late 1990s began with great promise but ended in spectacular failure, creating one of the most significant energy crises in American history. The California experiment became a cautionary tale that would fundamentally reshape how policymakers, regulators, and industry participants approached electricity market reform. Understanding the design flaws, market dynamics, and political pressures that led to California's crisis provides crucial lessons about the complexities of transforming regulated monopolies into competitive markets while maintaining system reliability and affordable service.

The Promise of Deregulation in the Golden State

California embarked on electricity deregulation with enormous optimism and ambition, driven by the belief that competition would deliver lower prices, improved efficiency, and greater innovation to the state's electricity customers. The state's economy was booming in the late 1990s, driven by the dot-com revolution and rapid growth in high-technology industries that were major electricity consumers. California's electricity rates were among the highest in the nation, creating strong political pressure to find ways to reduce costs for both residential and business customers.

The deregulation movement in California was supported by a diverse coalition that included large industrial customers seeking lower electricity costs, environmental groups hoping that competition would accelerate the adoption of cleaner technologies, and free-market advocates who believed that competitive forces would prove superior to traditional regulation. This broad political support enabled the passage of Assembly Bill 1890 in 1996, which established the framework for California's transition to competitive electricity markets.

The legislation promised immediate rate reductions for residential customers while creating competitive wholesale markets that would drive down costs over time. Proponents argued that California's high electricity rates were largely the result of inefficient regulated monopolies and poor regulatory decisions, particularly regarding nuclear power investments. They contended that competition would force utilities to operate more efficiently while encouraging new market entrants to build more cost-effective generating capacity.

The optimism surrounding California's deregulation effort was reinforced by successful deregulation experiences in other industries and other states. Pennsylvania had implemented electricity deregulation in 1996 with apparent success, achieving rate reductions and increased customer choice without major disruptions. The successful deregulation of telecommunications, airlines, and other industries provided additional evidence that competitive markets could deliver benefits that regulated monopolies could not achieve.

Flawed Market Design - The Seeds of Crisis

California's electricity market design contained fundamental flaws that would ultimately lead to its collapse, though these problems were not immediately apparent to policymakers and market participants. The state's approach was a hybrid model that attempted to combine elements of regulation and competition in ways that created perverse incentives and systemic vulnerabilities.

One of the most critical design flaws was the decision to force utilities to sell their generating assets while requiring them to purchase power for their customers through a centralized spot market called the California Power Exchange (PX). This arrangement prevented utilities from entering into long-term contracts that could have provided price stability and supply security. Instead, utilities had to purchase all their power on a day-to-day basis at prices that fluctuated dramatically based on supply and demand conditions.

The requirement to purchase power exclusively through the spot market created enormous financial risks for utilities while providing opportunities for generators to exercise market power. When supply became tight, generators could dramatically increase prices knowing that utilities had no alternatives and would be forced to pay whatever prices were necessary to meet their customers' demands. This market structure essentially transferred all commodity price risk to utilities while eliminating their ability to manage that risk through long-term contracting.

California's retail rate freeze, designed to provide immediate benefits to customers, created additional problems by preventing utilities from passing through wholesale price increases to customers. When wholesale prices spiked above retail rates, utilities faced massive financial losses that threatened their creditworthiness and ability to continue purchasing power. This rate freeze was intended to last only during the initial transition period, but it became politically difficult to remove as wholesale prices began rising.

The separation of energy and transmission services created coordination problems that had not existed under the integrated utility model. The California Independent System Operator (CAISO), responsible for managing the transmission system and maintaining reliability, had to coordinate with the Power Exchange and numerous market participants without the authority structures that had characterized integrated utility operations. This fragmentation of responsibility made it more difficult to anticipate and respond to system problems.

Market Manipulation and Gaming

The complex structure of California's electricity markets created numerous opportunities for market manipulation and gaming that sophisticated market participants exploited to increase their profits at the expense of California utilities and customers. These gaming strategies were often technically legal under existing market rules but violated the spirit of competition that deregulation was intended to promote.

One common gaming strategy involved creating artificial transmission congestion by strategically scheduling power flows that would create bottlenecks in the transmission system. When transmission lines became congested, generators located in areas with limited transmission capacity could demand premium prices for their power, since the system operator had few alternatives for meeting customer demand. Some companies deliberately created these congestion situations to increase their market power and capture higher prices.

"Round-trip" trading involved companies buying and selling the same power multiple times to create the appearance of market activity and manipulate price indices that were used to settle long-term contracts. These trades did not represent real economic activity but could influence market prices in ways that benefited the companies engaging in the trades while harming other market participants.

Generators also manipulated the market by withholding capacity during peak demand periods, creating artificial scarcity that drove up prices. Companies would report their power plants as unavailable for maintenance or technical reasons during times when high prices could be expected, then mysteriously find that the plants could operate when prices spiked high enough to make operation profitable. This strategy was particularly effective because the market had limited ability to distinguish between genuine technical problems and strategic manipulation.

The complexity of California's market rules made it difficult for regulators to detect and prevent these gaming strategies in real time. Market participants employed teams of analysts and traders who studied the market rules in detail to identify opportunities for profit that regulators and policymakers had not anticipated. The rapid pace of market transactions and the technical complexity of electricity systems made oversight challenging even when regulators suspected manipulation was occurring.

The Crisis Unfolds - 2000-2001

The California electricity crisis began to manifest in summer 2000 when a combination of hot weather, low hydroelectric generation, and high natural gas prices created tight supply conditions that exposed the vulnerabilities of the state's market design. As wholesale electricity prices began to spike, utilities found themselves paying dramatically more for power than they could recover through frozen retail rates, creating mounting financial losses that threatened their ability to continue operations.

The crisis escalated during the winter of 2000-2001 when continued high wholesale prices and growing utility financial distress led to concerns about the creditworthiness of California utilities. As their credit ratings deteriorated, suppliers became reluctant to sell power to the utilities, creating supply shortages that forced the system operator to implement rolling blackouts to maintain system stability. These blackouts, the first in California since World War II, shocked residents and businesses who had taken reliable electricity service for granted.

Pacific Gas & Electric and Southern California Edison, two of the state's largest utilities, accumulated billions of dollars in unrecovered power purchase costs as wholesale prices remained well above retail rates. The utilities' mounting debt and deteriorating financial condition created a crisis of confidence that made suppliers increasingly reluctant to enter into transactions with them, exacerbating supply shortages and price volatility.

The state government was forced to intervene directly in electricity markets when utility financial distress threatened to cause system-wide collapse. The California Department of Water Resources began purchasing power directly on behalf of the utilities, using state credit to ensure that suppliers would continue delivering electricity to California customers. This intervention represented a dramatic reversal of the deregulation philosophy and demonstrated the critical importance of maintaining creditworthy buyers in electricity markets.

Rolling blackouts became a regular occurrence during the winter of 2000-2001, affecting businesses, schools, and residential customers throughout the state. These interruptions caused economic losses estimated in the billions of dollars while undermining confidence in California's ability to maintain reliable electricity service. The blackouts also had significant political consequences, contributing to the eventual recall of Governor Gray Davis and highlighting the political risks associated with electricity market failures.

Economic and Social Consequences

The California electricity crisis had devastating economic consequences that extended far beyond the electricity sector itself. The state's economy, which had been growing rapidly during the dot-com boom, was severely affected by unreliable electricity service and rapidly rising energy costs. High-technology companies, which formed the backbone of California's economic growth, were particularly vulnerable to power interruptions that could cause costly production disruptions and data losses.

Manufacturing businesses throughout the state were forced to reduce production or shut down operations entirely during rolling blackouts, creating supply chain disruptions that affected companies throughout the country. Some businesses invested in backup generators or relocated operations to other states where electricity service was more reliable and affordable. The crisis undermined California's reputation as a business-friendly environment and contributed to concerns about the state's long-term economic competitiveness.

Residential customers faced dramatic increases in electricity bills as retail rate caps were eventually lifted and the costs of power purchases were passed through to consumers. Many families struggled to pay electricity bills that doubled or tripled from pre-crisis levels, creating financial hardship and political anger that lasted long after the crisis ended. Low-income customers were particularly affected, as they had fewer resources to absorb higher energy costs and limited ability to reduce their electricity consumption.

The crisis also had significant environmental consequences, as the state's focus on maintaining electricity supply led to the relaxation of environmental regulations and the operation of older, more polluting power plants that might otherwise have been retired. Air quality deteriorated in some areas as utilities and system operators prioritized supply reliability over environmental protection during emergency conditions.

Public confidence in electricity markets and deregulation was severely damaged by the crisis, creating political resistance to market-based approaches that persisted for years after the immediate problems were resolved. The crisis demonstrated that electricity markets were far more complex and fragile than many policymakers had understood, requiring careful design and ongoing oversight to function effectively.

Regulatory and Policy Responses

The federal response to the California electricity crisis involved multiple agencies and approaches as regulators struggled to understand and address the complex problems that had emerged. The Federal Energy Regulatory Commission (FERC) initially resisted calls for comprehensive price controls, arguing that market-based pricing was necessary to encourage new supply and demand response. However, as the crisis deepened,

FERC imposed "soft" price caps and other market interventions designed to prevent the most egregious forms of price manipulation.

FERC also launched extensive investigations into market manipulation and gaming strategies that had contributed to the crisis. These investigations revealed the extent to which market participants had exploited flaws in California's market design to increase their profits at the expense of California consumers. The agency eventually ordered hundreds of millions of dollars in refunds from generators and marketers who had charged excessive prices during the cris s period.

The California Public Ut lities Commission was forced to abandon key elements of the deregulation program as the crisis demonstrated their unworkabi ity. The requirement that utilities purchase all their power through the spot market was eliminated, allowing utilities to enter into long-term contracts that could provide greater price stability and supply security. Retail rate caps were eventually lifted, though the political process of passing through power costs to customers remained controversial and incomplete.

State legislation was enacted to provide greater oversight of electricity markets and prevent some of the gaming strategies that had contributed to the crisis. New market power mitigation measures were implemented to limit generators' ability to withhold capacity or manipulate prices during periods of tight supply. However, many observers argued that these reforms were insufficient to address the fundamental structural problems that had caused the crisis.

The crisis also led to renewed emphasis on energy efficiency and demand response programs as alternatives to building new generating capacity. California implemented aggressive efficiency standards and programs that helped reduce electricity demand growth and improve the balance between supply and demand. These programs became models for other states and helped establish California as a leader in energy efficiency policy.

Lessons Learned and Long-term Implications

The California electricity crisis provided numerous lessons about the challenges of designing and implementing competitive electricity markets. Perhaps the most important lesson was that market design details matter enormously - seemingly minor features of market rules and institutions can have major consequences for market performance and stability. The crisis demonstrated that electricity markets cannot be designed through theoretical models alone but require careful attention to practical implementation challenges and potential unintended consequences.

The crisis also highlighted the importance of maintaining adequate supply margins and investment incentives in competitive markets. California's market design provided insufficient incentives for new capacity investment while creating opportunities for existing generators to exercise market power during tight supply conditions. This experience influenced subsequent market designs that placed greater emphasis on capacity markets and long-term resource adequacy.

The role of financial intermediaries and credit support in electricity markets became apparent during the California crisis, as utility financial distress created supply shortages that threatened system reliability. This experience led to greater attention to credit requirements and financial responsibility in wholesale power markets, with new rules designed to prevent financially weak participants from threatening market stability.

The California crisis also demonstrated the political risks associated with electricity market failures, as the economic and social disruptions created lasting damage to public support for deregulation. The crisis showed that electricity is too important to the economy and society to be treated as an ordinary commodity, requiring market designs that prioritize reliability and affordability alongside efficiency and competition.

The End of an Era

By 2002, California had largely abandoned its deregulation experiment, returning to a more traditional regulatory model with some competitive elements. The state's utilities resumed owning generating capacity and entering into long-term contracts, while retail competition was suspended indefinitely. The crisis had cost the state an estimated $40 billion to $45 billion in higher electricity costs and economic losses, making it one of the most expensive policy failures in American history.

The California experience effectively ended the momentum toward nationwide electricity deregulation that had been building throughout the 1990s. Many states that had been considering deregulation postponed or abandoned their plans, while others implemented more cautious approaches that attempted to avoid the pitfalls that had trapped California. The crisis demonstrated that while competitive electricity markets could potentially deliver benefits, they required much more careful design and implementation than had been initially understood.

Chapter 17: The Enron Scandal - Corruption and Market Manipulation

The collapse of Enron Corporation in 2001 sent shockwaves throughout the American business world and fundamentally altered perceptions of corporate governance, financial markets, and energy industry regulation. What had been celebrated as one of the most innovative and successful companies of the 1990s was revealed to be built on a foundation of accounting fraud, market manipulation, and corporate corruption that ultimately destroyed the company and caused billions of dollars in losses for investors, employees, and consumers. The Enron scandal had particular significance for the electrical industry because the company had been a major player in power markets and its manipulative practices had contributed significantly to the California electricity crisis.

The Rise of Enron - Innovation or Illusion

Enron's transformation from a traditional natural gas pipeline company to a high-flying energy trading and services conglomerate appeared to represent the future of the energy industry in the deregulated world of the 1990s. The company positioned itself as a "New Economy" enterprise that used sophisticated financial instruments, information technology, and market-making capabilities to create value in ways that traditional energy companies could not match.

Founded in 1985 through the merger of Houston Natural Gas and InterNorth, Enron initially operated as a conventional natural gas pipeline company under the leadership of Kenneth Lay. However, the company began to transform itself in the early

1990s under the influence of Jeffrey Skilling, a former McKinsey & Company consultant who brought Wall Street financial techniques to the energy industry. Skilling advocated for Enron to become an "asset-light" company that would generate profits through trading and market-making rather than owning and operating physical energy infrastructure.

The deregulation of natural gas markets provided Enron with its first major opportunity to implement this strategy. As gas prices were deregulated and interstate pipeline transportation was opened to competition, Enron developed expertise in gas trading and marketing that enabled it to profit from price differences and supply imbalances. The company's success in gas markets provided the foundation for expansion into electricity, water, broadband, and other commodity markets.

Enron's business model emphasized financial innovation and risk management services that promised to help customers manage the price volatility and supply risks associated with deregulated energy markets. The company developed complex derivative instruments and structured transactions that were designed to transfer risks between different market participants while generating fee income for Enron. This approach appeared to create enormous value with relatively little capital investment, generating impressive returns for shareholders and massive compensation for executives.

The company's rapid growth and apparent profitability made it a darling of Wall Street analysts and business media, who praised Enron's innovative approach and predicted continued success as energy markets became increasingly competitive. Fortune magazine named Enron "America's Most Innovative Company" for six consecutive years from 1995 to 2000, while business schools used Enron as a case study of successful corporate transformation and strategic innovation.

Market Manipulation and Gaming Strategies

Behind Enron's public image of innovation and success lay a systematic pattern of market manipulation and gaming that exploited regulatory gaps and market design flaws to generate artificial profits at the expense of consumers and market integrity. The company's trading operations developed increasingly sophisticated strategies for manipulating electricity markets, particularly in California where market design flaws created numerous opportunities for profitable gaming.

Enron's California trading strategies were given colorful names that reflected their manipulative nature: "Death Star," "Get Shorty," "Fat Boy," and "Ricochet" were among the schemes that Enron traders used to extract money from California's electricity markets. These strategies involved creating artificial congestion on transmission lines, manipulating day-ahead and real-time market prices, and exploiting differences between California's market rules and neighboring systems.

The "Death Star" strategy involved scheduling power flows on transmission lines to create artificial congestion, then collecting congestion payments from the system operator for relieving the congestion that Enron had artificially created. This strategy generated millions of dollars in payments from California ratepayers without providing any real economic value or improving system reliability. The scheme worked because California's market rules automatically paid for congestion relief without adequately verifying that the congestion was genuine.

"Fat Boy" involved over-scheduling load in California to receive payments for demand reduction, even when actual customer demand had not increased. Enron would submit inflated demand forecasts to the California system operator, then receive payments for reducing the fictitious excess demand. This strategy exploited the system operator's need to maintain supply-demand balance and willingness to pay for demand reductions during tight supply conditions.

These gaming strategies were profitable for Enron but imposed significant costs on California electricity customers and undermined the integrity of the state's electricity markets. The company's traders celebrated their ability to exploit market flaws and extract money from California, with recorded phone conversations revealing their awareness that their strategies were harming California consumers and contributing to the state's electricity crisis.

Accounting Fraud and Financial Engineering

Enron's market manipulation was accompanied by increasingly aggressive accounting practices that obscured the company's true financial condition while inflating reported earnings and revenue. The company used a complex web of special purpose entities (SPEs) and off-balance-sheet partnerships to hide debt, inflate profits, and avoid regulatory oversight that might have revealed the extent of its manipulation and financial engineering.

The company's use of mark-to-market accounting for its trading operations enabled it to book anticipated future profits immediately upon entering into long-term contracts, even when those profits might never materialize. This accounting treatment allowed Enron to report steady earnings growth that impressed Wall Street analysts and investors, even as the company's actual cash flows became increasingly problematic.

Enron's partnership with Arthur Andersen, one of the "Big Five" accounting firms, facilitated the company's aggressive accounting practices through inadequate auditing and consulting relationships that created conflicts of interest. Andersen's Houston office developed particularly close relationships with Enron executives and appeared to prioritize client satisfaction over accounting integrity, approving questionable transactions and accounting treatments that inflated Enron's reported financial performance.

The company's financial engineering became increasingly complex and risky as executives sought to maintain the appearance of continued growth and profitability. Enron created hundreds of subsidiaries and partnerships, many with names like "Jedi," "Chewco," and "Raptor" that reflected the Star Wars obsessions of company executives. These entities were used to shift debt off Enron's balance sheet, manipulate earnings, and enrich senior executives through undisclosed compensation arrangements.

Chief Financial Officer Andrew Fastow developed particularly sophisticated schemes involving partnerships that he controlled personally, creating conflicts of interest that enabled him to extract millions of dollars from Enron while hiding the company's mounting debt and declining profitability. These partnerships were designed to circumvent accounting rules and regulatory requirements while providing Fastow and other executives with enormous personal profits.

Corporate Culture and Executive Behavior

Enron's corporate culture played a crucial role in enabling the fraud and manipulation that ultimately destroyed the company. The company promoted a hyper-competitive, results-oriented environment that emphasized individual performance and financial rewards while downplaying ethical considerations and long-term sustainability. This culture attracted ambitious individuals who were willing to bend or break rules to achieve short-term success.

The company's compensation system provided enormous incentives for executives and traders to generate short-term profits regardless of the means used to achieve them. Annual bonuses could reach millions of dollars for successful traders and executives, while stock options provided additional incentives to inflate short-term stock prices through whatever means necessary. This compensation structure encouraged risk-taking and rule-bending while discouraging questions about the sustainability or legality of the company's practices.

Enron's "rank and yank" performance evaluation system, which required managers to annually fire the bottom 10 percent of their employees, created additional pressure for individuals to achieve results at any cost. This system discouraged collaboration and ethical behavior while rewarding individuals who could generate profits through any means available, including market manipulation and accounting fraud.

Senior executives, particularly CEO Jeffrey Skilling and Chairman Kenneth Lay, set a tone that prioritized financial performance over ethical behavior and regulatory compliance. Skilling in particular was known for his aggressive and confrontational style, dismissing critics and skeptics while promoting a vision of Enron as a revolutionary company that was creating new ways of doing business. This leadership style discouraged internal dissent and critical analysis of the company's practices.

The company's board of directors failed to provide adequate oversight of management and actively enabled many of the questionable practices that led to Enron's collapse. Board members with extensive business experience and impressive credentials approved transactions and compensation arrangements that clearly benefited executives at the expense of shareholders and other stakeholders. The board's failure to fulfill its oversight responsibilities was later cited as a prime example of corporate governance failures that required regulatory reform.

The Unraveling and Collapse

Enron's house of cards began to collapse in October 2001 when the company announced a $618 million third-quarter loss and a $1.2 billion reduction in shareholder equity related to partnerships controlled by CFO Andrew Fastow. This announcement raised questions about the company's financial practices and prompted investigations by regulators, analysts, and journalists who began to uncover the extent of Enron's fraud and manipulation.

The Securities and Exchange Commission launched a formal investigation into Enron's financial practices, while credit rating agencies began to downgrade the company's debt as questions arose about its true financial condition. As Enron's credit ratings fell, the company faced immediate demands for cash collateral from trading counterparties and lenders, creating a liquidity crisis that exposed the fragility of its financial structure.

Dynegy, a smaller energy trading company, initially agreed to acquire Enron for $9 billion in November 2001, but the deal collapsed as the full extent of Enron's problems became apparent. The failed merger attempt accelerated Enron's decline as investors and trading partners lost confidence in the company's ability to survive as an independent entity.

On December 2, 2001, Enron filed for bankruptcy protection, making it the largest corporate bankruptcy in American history at that time. The bankruptcy filing revealed the extent of the company's debt and the complexity of its corporate structure, with over 3,000 subsidiaries and partnerships around the world. Thousands of employees lost their jobs and retirement savings, while investors lost billions of dollars in stock and bond investments.

The collapse of Arthur Andersen, Enron's accounting firm, followed shortly after as the firm was indicted for obstruction of justice related to its destruction of Enron-related documents. Andersen's conviction effectively ended the firm's operations, eliminating one of the world's largest accounting firms and reducing the "Big Five" accounting firms to the "Big Four" that continue to dominate the accounting industry today.

Impact on Energy Markets and Regulation

The revelation of Enron's market manipulation practices had significant implications for energy market regulation and design, particularly in wholesale electricity markets where the company had been a major participant. Federal regulators were forced to acknowledge that existing market rules and oversight mechanisms were inadequate to prevent sophisticated manipulation strategies that harmed consumers and undermined market integrity.

FERC launched comprehensive investigations into Enron's trading practices and eventually ordered the company to make refunds totaling over $1.5 billion to customers who had been harmed by its manipulation strategies. However, collecting these refunds proved difficult due to Enron's bankruptcy, leaving many consumers without adequate compensation for the damages they had suffered.

The scandal accelerated reforms in electricity market design and oversight that were already being considered in response to the California electricity crisis. New market monitoring and mitigation measures were implemented to detect and prevent manipulation strategies similar to those employed by Enron. Market rules were modified to close loopholes that had enabled gaming and manipulation, while penalty structures were enhanced to deter future misconduct.

The Enron scandal also led to increased skepticism about the benefits of electricity deregulation and market-based approaches to energy policy. The revelation that one of the most prominent advocates and beneficiaries of deregulation had been systematically manipulating markets undermined public confidence in competitive energy markets and slowed the momentum toward further deregulation.

Federal oversight of energy markets was significantly enhanced through new legislation and regulatory initiatives designed to prevent future market manipulation and fraud. The Energy Policy Act of 2005 gave FERC new authorities to investigate and penalize market manipulation, while the Commodity Futures Trading Commission received enhanced authority over energy derivatives markets.

Legal Consequences and Criminal Prosecutions

The Enron scandal resulted in extensive criminal prosecutions that sent several senior executives to prison while establishing important precedents for corporate accountability and white-collar crime enforcement. The Department of Justice created an Enron Task Force that pursued charges against over 30 individuals associated with the company's fraud and manipulation schemes.

Jeffrey Skilling, Enron's former CEO, was convicted on 19 counts of conspiracy, fraud, and insider trading and sentenced to 24 years in prison (later reduced to 14 years). Kenneth Lay, the company's founder and chairman, was convicted on six counts of conspiracy and fraud but died of a heart attack before sentencing, resulting in the vacation of his conviction under federal law.

Andrew Fastow, the CFO who created many of Enron's fraudulent partnerships, pleaded guilty to two counts of conspiracy and was sentenced to six years in prison after cooperating with prosecutors in the cases against other executives. Several other senior executives, including Chief Accounting Officer Richard Causey and Treasurer Ben Glisan, also pleaded guilty and received prison sentences.

The prosecutions established important precedents for holding senior executives personally accountable for corporate fraud, even when they claimed ignorance of specific illegal activities. The cases demonstrated that executives could not insulate themselves from criminal liability by delegating responsibility for questionable practices to subordinates while continuing to benefit from the results of those practices.

Civil lawsuits filed by shareholders, employees, and other parties harmed by Enron's collapse resulted in settlements totaling billions of dollars, though actual recoveries were limited by the company's bankruptcy. These lawsuits established precedents for holding accountants, lawyers, and financial institutions accountable for their roles in enabling corporate fraud, leading to significant changes in professional practices and standards.

Lasting Legacy and Lessons Learned

The Enron scandal fundamentally altered American corporate governance, financial reporting, and regulatory oversight in ways that continue to influence business practices today. The scandal demonstrated the importance of ethical leadership, independent oversight, and transparent financial reporting in maintaining public confidence in corporations and financial markets.

The passage of the Sarbanes-Oxley Act in 2002 established new standards for corporate accountability and financial reporting that were directly influenced by the lessons of the Enron scandal. This legislation enhanced penalties for corporate fraud, improved auditor independence, and required senior executives to personally certify their companies' financial statements, creating new legal risks for executives who might be tempted to engage in fraudulent practices.

The scandal also led to significant reforms in energy market regulation and oversight that improved the integrity and efficiency of wholesale electricity markets. New market monitoring capabilities, enhanced penalty structures, and improved coordination between federal and state regulators helped prevent the types of manipulation that Enron had pioneered in California and other markets.

Perhaps most importantly, the Enron scandal served as a reminder that sophisticated financial engineering and complex corporate structures could be used to hide fundamental business problems and ethical failures that would eventually surface with devastating consequences. The scandal reinforced the importance of transparency, accountability, and ethical behavior in business while demonstrating the costs that fraud and manipulation impose on society as a whole.

Chapter 18: Re-regulation and the Aftermath of Enron

The collapse of Enron and the California electricity crisis fundamentally transformed the landscape of American energy policy and utility regulation, ushering in an era of re-regulation that sought to address the failures and excesses of the deregulation movement while preserving the benefits that competition could provide. This period, spanning from 2002 to approximately 2010, was characterized by enhanced regulatory oversight, stricter corporate governance requirements, and a more cautious approach to market-based energy policies. The aftermath of Enron forced policymakers, regulators, and industry participants to reconsider fundamental assumptions about the role of markets in the energy sector while developing new frameworks for balancing competition with consumer protection and system reliability.

The Sarbanes-Oxley Response

The most immediate and comprehensive response to the Enron scandal was the passage of the Sarbanes-Oxley Act of 2002, landmark legislation that established new standards for corporate governance, financial reporting, and executive accountability across all American public companies. While Sarbanes-Oxley applied to all industries, its impact on energy companies was particularly significant given the sector's prominence in the corporate scandals that had triggered the legislation.

The act's requirements for CEO and CFO certification of financial statements created new personal liability for energy company executives who had previously been insulated from direct responsibility for accounting irregularities. Under Section

302 of the act, senior executives had to personally certify that their companies' quarterly and annual reports did not contain material misstatements and fairly represented the company's financial condition. This requirement fundamentally altered the relationship between executives and their financial reporting, creating strong incentives for more careful oversight of accounting practices.

Section 404 of Sarbanes-Oxley required companies to establish and maintain internal controls over financial reporting, with independent auditors required to attest to the effectiveness of these controls. For energy companies with complex trading operations and multiple subsidiaries, this requirement necessitated substantial investments in compliance systems and procedures that increased operating costs while improving transparency and accountability.

The act also enhanced penalties for corporate fraud, with maximum prison sentences increased to 25 years for securities fraud and 20 years for mail and wire fraud. These enhanced penalties were specifically designed to deter the types of executive misconduct that had characterized the Enron scandal, creating real personal consequences for executives who might be tempted to engage in fraudulent practices.

Sarbanes-Oxley's provisions for auditor independence addressed the conflicts of interest that had enabled Arthur Andersen's complicity in Enron's fraud. The act prohibited accounting firms from providing certain consulting services to their audit clients while requiring audit partner rotation every five years. These requirements were designed to ensure that auditors would maintain professional skepticism and independence from their clients rather than becoming too closely aligned with management interests.

Enhanced Energy Market Oversight

The Federal Energy Regulatory Commission (FERC) responded to the Enron scandal by significantly enhancing its oversight of wholesale energy markets and its authority to investigate and punish market manipulation. The commission recognized that its existing regulatory framework had been inadequate to detect and prevent the sophisticated gaming strategies that Enron and other companies had employed to manipulate electricity markets.

FERC established new market monitoring units within each Regional Transmission Organization (RTO) and Independent System Operator (ISO) to provide real-time surveillance of wholesale electricity markets. These market monitors were granted broad authority to investigate suspicious trading patterns, analyze market outcomes, and recommend enforcement actions when manipulation or abuse was detected. The monitoring systems used sophisticated analytical tools to identify unusual price patterns, bidding behavior, or trading strategies that might indicate market manipulation.

The commission also enhanced its enforcement capabilities by creating the Office of Enforcement as a separate division within FERC with dedicated staff and resources for investigating market manipulation and other violations of federal energy law. This organizational change reflected FERC's recognition that effective market oversight required specialized expertise and dedicated resources rather than treating enforcement as a secondary responsibility.

New market rules were implemented across all wholesale electricity markets to address the specific gaming strategies that had been employed during the California crisis and exposed by the Enron investigations. These rules included enhanced penalties for market manipulation, stricter requirements for generator availability reporting, and improved coordination between day-ahead and real-time markets to reduce opportunities for gaming.

FERC also expanded its authority over energy market participants through broader definitions of market manipulation and enhanced information-gathering powers. The commission gained the ability to require market participants to provide detailed information about their trading strategies, financial positions, and corporate relationships that could affect market outcomes. These enhanced authorities provided FERC with better tools for detecting and preventing market manipulation before it could cause significant harm to consumers.

Corporate Governance Reforms in the Energy Sector

The energy industry implemented significant corporate governance reforms that went beyond the minimum requirements of Sarbanes-Oxley, responding to investor and public demands for enhanced accountability and transparency. These reforms were driven partly by regulatory requirements but also by recognition that strong governance practices were essential for maintaining access to capital markets and public confidence.

Many energy companies restructured their boards of directors to enhance independence and expertise, increasing the proportion of independent directors and establishing specialized committees for audit, compensation, and risk management.

Board committees gained enhanced authority and responsibility for overseeing management practices that had contributed to the scandals of the early 2000s.

Risk management practices were significantly enhanced across the energy industry as companies recognized that inadequate risk oversight had contributed to many of the problems that had emerged during the deregulation period. New risk management frameworks were implemented that provided better visibility into trading positions, market exposures, and operational risks that could affect company performance.

Executive compensation practices were reformed to reduce incentives for short-term risk-taking and manipulation while better aligning management interests with long-term shareholder value. Many companies eliminated or reduced the emphasis on short-term performance metrics in executive compensation while implementing claw back provisions that allowed boards to recover compensation based on restated financial results.

Internal audit functions were significantly enhanced to provide independent oversight of company operations and compliance with new regulatory requirements. Many companies expanded their internal audit staff and enhanced their authority to investigate potential problems without management interference. These enhanced internal audit functions provided early warning systems for detecting problems before they could threaten company viability or result in regulatory sanctions.

The Slowdown of Deregulation

The Enron scandal and California crisis effectively ended the momentum toward nationwide electricity deregulation that had been building throughout the 1990s. States that had been considering deregulation postponed or abandoned their plans, while others implemented more limited and cautious approaches that attempted to avoid the pitfalls that had led to problems in California and other early deregulation states.

By 2005, only 17 states plus the District of Columbia had implemented some form of retail electricity competition, compared to earlier predictions that most states would eventually adopt deregulated market structures. Several states that had initially moved toward deregulation, including California, Montana, and Nevada, suspended or reversed their deregulation programs following the crises of the early 2000s.

The states that continued with deregulation implemented more conservative market designs that emphasized consumer protection and system reliability over pure market efficiency. These revised approaches included stronger consumer protections, more robust market monitoring, and greater regulatory oversight of market participants. The emphasis shifted from rapid implementation of competitive markets to careful design and gradual implementation that could avoid the problems experienced in earlier deregulation efforts.

Federal policy also became more cautious about promoting deregulation, with the Energy Policy Act of 2005 including enhanced penalties for market manipulation while repealing the Public Utility Holding Company Act of 1935. This mixed approach reflected the recognition that while competitive markets could provide benefits, they required careful regulation and oversight to function effectively.

The investment community also became more skeptical of deregulated energy companies following the collapse of Enron and other energy traders. Credit ratings agencies implemented more stringent requirements for energy companies, particularly those with significant trading operations or merchant generation assets. This increased scrutiny made it more expensive for deregulated energy companies to access capital markets while favoring more conservative business models.

Technological and Operational Improvements

The period following the Enron scandal saw significant investments in technology and operational improvements designed to enhance market transparency, system reliability, and regulatory compliance. These improvements were driven partly by regulatory requirements but also by industry recognition that better systems and procedures were essential for maintaining public confidence and avoiding future crises.

Trading and risk management systems were significantly enhanced to provide better real-time monitoring of market positions and exposures. Energy companies invested heavily in sophisticated software systems that could track complex trading strategies, calculate risk exposures, and provide early warning of potential problems. These systems also improved regulatory compliance by providing detailed records of trading activities that could be reviewed by regulators and auditors.

Market information systems were improved to provide greater transparency about electricity market operations and outcomes. RTOs and ISOs implemented enhanced websites and data dissemination systems that provided market participants and regulators with better visibility into market conditions, price formation, and system operations. This improved transparency made it more difficult for market participants to engage in manipulation while helping regulators identify potential problems more quickly.

Grid monitoring and control systems were enhanced to improve system reliability and reduce the potential for market manipulation through physical system operations. Investments in advanced metering, communication systems, and automated controls provided better real-time visibility into system conditions while enabling faster response to emerging problems.

Cybersecurity became a new focus area for energy companies and regulators as the increased reliance on electronic systems and data communications created new vulnerabilities that could be exploited by malicious actors. The energy industry began implementing comprehensive cybersecurity programs designed to protect critical infrastructure from both external attacks and internal threats.

Financial Market Reforms

The energy derivatives markets that had been central to Enron's trading operations were subjected to enhanced oversight and regulation designed to prevent manipulation and improve transparency. The Commodity Futures Trading Commission (CFTC) gained enhanced authority over energy derivatives markets while implementing new reporting requirements and position limits designed to prevent excessive speculation and manipulation.

Credit requirements for energy market participants were significantly enhanced to reduce the risk of financial collapse that could threaten market stability. Clearinghouses and exchanges implemented higher margin requirements and more stringent credit standards for energy trading, while market participants were required to provide more detailed financial information and collateral to support their trading activities.

The over-the-counter (OTC) energy derivatives markets were subjected to greater oversight and standardization following the implementation of the Dodd-Frank Act in 2010. While Dodd-Frank was primarily a response to the financial crisis of 2008, its provisions for derivatives market regulation also addressed concerns about energy market manipulation and systemic risk that had been highlighted by the Enron scandal.

Bank lending practices for energy companies were reformed to reduce the risks associated with complex corporate structures and aggressive financial engineering. Banks implemented enhanced due diligence procedures for energy loans while requiring more detailed disclosure of trading activities, risk management practices, and corporate governance arrangements.

Long-term Industry Transformation

The re-regulation period following the Enron scandal resulted in fundamental changes to the energy industry that continue to influence its operations today. The enhanced regulatory oversight and corporate governance requirements established new standards for industry behavior that raised the cost of doing business while improving transparency and accountability.

Many energy companies simplified their corporate structures and business models in response to investor and regulatory demands for greater transparency and focus. Complex trading operations and speculative investments were reduced or eliminated in favor of more traditional utility operations that could be more easily understood and regulated.

The consolidation of the energy industry accelerated during this period as smaller companies found it difficult to bear the costs of enhanced compliance and regulatory oversight. Larger companies with greater resources were better positioned to implement the systems and procedures required by new regulations while spreading these costs across larger customer bases.

Public attitudes toward the energy industry were permanently altered by the scandals of the early 2000s, with consumers and policymakers maintaining greater skepticism about utility practices and market-based energy policies. This skepticism influenced energy policy debates for years after the immediate crises had passed, contributing to more cautious approaches to energy market reform and utility regulation.

Setting the Stage for Modern Energy Markets

By 2010, the American energy industry had stabilized following the turbulent period that began with the California crisis and culminated in the Enron scandal. The enhanced regulatory framework and corporate governance practices that emerged from this period provided a more stable foundation for energy markets while maintaining many of the competitive elements that had been introduced during the deregulation movement.

The lessons learned from the re-regulation period would prove valuable as the energy industry faced new challenges related to renewable energy integration, climate change, and technological innovation. The enhanced oversight and transparency requirements established during this period provided tools for managing these new challenges while maintaining system reliability and consumer protection.

The balance between competition and regulation that emerged from the re-regulation period represented a more mature approach to energy market design that recognized both the benefits and limitations of market-based approaches. This balanced approach would continue to influence energy policy development as the industry evolved to address the environmental and technological challenges of the twenty-first century.

Chapter 19: Modern Regulated Utilities - Balancing Reliability and Cost

Following the turbulent period of deregulation experiments and the Enron scandal, a significant portion of the American electrical industry returned to or remained under traditional regulated utility models, albeit with important modifications informed by the lessons learned from market-based approaches. Today's regulated utilities operate in a fundamentally different environment than their predecessors, facing unprecedented challenges from aging infrastructure, environmental regulations, technological change, and evolving customer expectations while maintaining their core obligation to provide reliable, affordable electrical service to all customers within their service territories.

The Evolution of Modern Regulation

The regulatory framework governing modern utilities has evolved significantly from the traditional cost-of-service model that dominated the industry through the mid-twentieth century. While the basic principle remains the same - utilities are guaranteed recovery of prudently incurred costs plus a reasonable return on investment - the implementation has become more sophisticated and performance-oriented in response to lessons learned from both successful and failed deregulation experiments.

State utility commissions, which retain primary authority over retail electricity regulation in most of the United States, have adopted more nuanced approaches that attempt to balance the consumer protection benefits of regulation with incentives for efficiency and innovation that competitive markets can provide.

Performance-based ratemaking has become increasingly common, with utilities' returns tied to their achievement of specific performance metrics related to reliability, customer satisfaction, environmental performance, or cost management.

Modern regulatory approaches also place greater emphasis on integrated resource planning, requiring utilities to demonstrate that their investment and operational decisions serve the long-term interests of customers and society rather than simply maximizing utility profits. These planning processes must consider not only traditional factors like load growth and system reliability but also environmental impacts, renewable energy integration, demand-side management opportunities, and emerging technologies that could affect future system needs.

The regulatory compact between utilities and their regulators has also evolved to address concerns about utility incentives that were highlighted during the deregulation period. Traditional regulation created incentives for utilities to over-invest in capital projects (the Averch-Johnson effect) while providing limited rewards for operational efficiency or innovation. Modern regulatory frameworks attempt to align utility incentives with customer interests through mechanisms like performance incentives, shared savings programs, and accelerated cost recovery for beneficial investments.

State Utility Commissions in the Modern Era

State public utility commissions continue to serve as the primary regulatory authorities for most American electricity customers, though their roles and responsibilities have expanded significantly beyond the traditional rate-setting function. Modern utility commissions must balance multiple, sometimes conflicting objectives including affordability, reliability, environmental protection, economic development, and technological innovation while maintaining their fundamental responsibility to protect consumers from potential monopoly abuse.

The complexity of modern utility regulation requires commissioners and staff with sophisticated technical and economic expertise spanning areas that were not traditionally within utility commission purview. Commissioners must understand renewable energy technologies, energy storage systems, demand response programs, cybersecurity threats, and climate change impacts while maintaining expertise in traditional areas like engineering economics, accounting, and legal precedent.

Many state commissions have adopted more collaborative and forward-looking approaches to regulation that engage stakeholders in long-term planning processes rather than simply responding to utility rate requests. These approaches recognize that the rapid pace of technological and environmental change requires proactive planning and stakeholder engagement to ensure that regulatory decisions serve long-term public interests rather than short-term political or economic pressures.

The geographic and political diversity of state regulation has created a natural laboratory for different regulatory approaches, with some states emphasizing market-oriented solutions while others maintain more traditional command-and-control approaches. This diversity enables useful comparisons of different regulatory strategies while allowing states to tailor their approaches to local conditions and preferences.

However, the state-based regulatory structure also creates challenges for addressing issues that transcend state boundaries, such as interstate transmission planning, regional reliability coordination, and climate change mitigation. The tension between state and federal authority over electricity regulation continues to evolve as the industry grapples with challenges that require coordination across multiple jurisdictions.

Infrastructure Challenges and Investment Needs

Modern regulated utilities face unprecedented infrastructure challenges that require massive capital investments while maintaining affordabil ty and reliability for customers. Over 70 percent of transmission lines are more than 25 years old and approaching the end of their typical 50-80 year lifecycle, creating urgent needs for replacement and upgrading that must be balanced against customer affordability concerns.

The aging infrastructure problem is compounded by changing usage patterns anc performance expectations that differ significantly from conditions when existing systems were designed and built. Original infrastructure was designed for one-way power flows from large central power plants to passive customers, but modern grids must accommodate two-way power flows from distributed energy resources, e ectric vehicle charging, and other new applications that stress systems in unexpected ways.

Climate change creates additional infrastructure challenges as extreme weather events become more frequent and severe, requiring utilities to nvest in resilience measures that can maintain service during hurricanes, wildfires, ice storms, and other natural disasters. These resilience investments often provide limited direct customer benefits during normal operations but are essential for maintaining service during emergency conditions that are becoming increasingly common.

Cybersecurity has emerged as a critical infrastructure concern as utilities become increasingly dependent on digital systems and communication networks that can be vulnerable to cyberattacks. Protecting critical infrastructure from both state-sponsored and criminal cyber threats requires ongoing investments in security systems, employee training, and emergency response capabilities that add to utility costs while providing benefits that customers may not directly perceive.

The integration of renewable energy sources creates additional infrastructure investment needs as traditional transmission and distribution systems must be modified to accommodate variable generation sources that can stress the grid in new ways. These modifications often require upgrades to protection systems, control equipment, and communication networks that enable the real-time coordination necessary for reliable operation with high levels of renewable penetration.

Environmental Regulations and Compliance Costs

Environmental regulations represent one of the most significant drivers of utility investment and operational changes in the modern era, fundamentally altering how utilities plan, build, and operate their systems. The transition from coal-fired generation to cleaner alternatives has required massive capital investments in new generating capacity while creating stranded asset risks for existing facilities that may become uneconomical before the end of their useful lives.

The Clean Air Act and its various amendments have established increasingly stringent requirements for air pollutant emissions from power plants, forcing utilities to install expensive pollution control equipment or retire older, less efficient facilities. These regulations have been particularly challenging for coal-fired power plants, many of which have been retired rather than upgraded due to the high costs of compliance relative to alternatives like natural gas and renewable energy.

State renewable portfolio standards and clean energy mandates require utilities to procure increasing percentages of their electricity from renewable sources, creating new planning and operational challenges while driving investments in wind, solar, and other clean energy technologies. These requirements often conflict with traditional utility preferences for dispatchable baseload generation, requiring new approaches to system planning and operation.

Carbon pricing policies and regulations, where implemented, create additional economic pressures that favor low-carbon generation sources while penalizing fossil fuel facilities. Even where carbon pricing has not been formally implemented, the prospect of future carbon regulations creates uncertainty that affects utility investment decisions and long-term planning processes.

Water quality and quantity regulations also affect utility operations, particularly for power plants that use significant amounts of water for cooling purposes. Regulations governing thermal discharge, water intake structures, and water consumption can require expensive modifications to existing facilities or influence decisions about new plant locations and technologies.

Technology Integration and Grid Modernization

The integration of new technologies represents both an opportunity and a challenge for modern regulated utilities, requiring substantial investments in grid modernization while potentially disrupting traditional utility business models. Smart grid technologies enable more efficient system operation and better customer service but require upfront investments that may take years to generate measurable benefits.

Advanced metering infrastructure (AMI) has been widely deployed by regulated utilities, providing customers with detailed usage information while enabling utilities to detect outages more quickly and operate their systems more efficiently. However, AMI deployments have been controversial in some jurisdictions due to customer privacy concerns and questions about cost-effectiveness relative to traditional metering systems.

Distributed energy resources, including rooftop solar, energy storage, and electric vehicles, create both opportunities and challenges for regulated utilities. While these resources can provide system benefits by reducing peak demand and improving resilience, they also reduce utility sales and can create power quality and safety issues if not properly managed. Utilities must develop new capabilities for managing these distributed resources while maintaining system reliability.

Energy storage technologies are increasingly being deployed by regulated utilities to provide grid services like frequency regulation, peak load management, and renewable energy integration. However, storage technologies remain expensive and their optimal applications are still being determined through pilot projects and demonstration programs that require regulatory approval and cost recovery mechanisms.

Electric vehicle adoption is creating new demands on the electrical system while providing potential opportunities for utilities to use vehicle batteries as distributed storage resources. Managing the grid impacts of widespread electric vehicle adoption requires new planning approaches, rate designs, and infrastructure investments that must be coordinated with transportation and environmental policies.

Customer Expectations and Engagement

Modern utility customers have fundamentally different expectations than previous generations, driven by experiences with competitive markets in other industries, environmental consciousness, and technological capabilities that enable greater control over energy consumption. These changing expectations require utilities to develop new capabilities and service offerings while maintaining their core reliability and affordability obligations.

Customer choice and customization have become increasingly important as customers seek greater control over their energy sources, usage patterns, and service options. While retail competition is not available in most regulated jurisdictions, utilities are finding ways to provide customers with choices regarding renewable energy, energy efficiency programs, and rate structures that can meet diverse customer preferences.

Digital engagement and self-service capabilities are now expected by customers who are accustomed to managing their financial, commercial, and personal affairs through online platforms and mobile applications. Utilities have invested heavily in customer information systems, mobile apps, and online portals that enable customers to monitor their usage, pay bills, report outages, and access energy efficiency programs.

Transparency and communication have become critical as customers demand better information about utility operations, rates, and policies. Social media and online communications enable rapid dissemination of information and customer feedback, requiring utilities to be more responsive and proactive in their customer communications than was previously necessary.

Environmental stewardship is increasingly important to customers who want their electricity to come from clean sources and their utilities to be responsible environmental actors. This customer preference has reinforced regulatory requirements for renewable energy and environmental protection while creating market opportunities for utilities that can demonstrate environmental leadership.

Financial Challenges and Business Model Evolution

The traditional utility business model, based on selling increasing quantities of electricity to cover fixed costs and earn regulated returns, faces significant challenges from technological change, environmental regulations, and evolving customer preferences. Sales growth has slowed or reversed in many jurisdictions due to energy efficiency improvements, distributed generation, and economic changes that reduce electricity consumption per capita.

Declining sales create financial pressures for utilities with high fixed costs, as lower volumes must cover the same infrastructure and operating expenses while maintaining service quality and reliability. This challenge has led to discussions about alternative rate designs and utility business models that can better align utility revenues with the costs they incur regardless of sales volumes.

Stranded asset risks have become a significant concern as environmental regulations and market forces accelerate the retirement of coal-fired power plants and other facilities that may not reach the end of their useful lives. Utilities and regulators must determine how to recover the remaining investments in these facilities while maintaining customer affordability and system reliability.

Access to capital markets remains critical for utilities that require substantial ongoing investments in infrastructure replacement and modernization. Maintaining strong credit ratings and investor confidence requires demonstrating that regulatory frameworks provide adequate cost recovery and returns while managing the risks associated with technological and regulatory change.

The Future of Regulated Utilities

The regulated utility model continues to evolve in response to technological, environmental, and economic pressures while maintaining its fundamental commitment to universal service and customer protection. The future likely involves hybrid approaches that combine traditional regulation with performance incentives, market mechanisms, and customer choice options that can deliver the benefits of competition while preserving the consumer protections and system reliability that regulation provides.

Regulatory innovation will continue to be essential as utilities adapt to changing conditions and technologies that were not contemplated under traditional regulatory frameworks. Performance-based ratemaking, multi-year rate plans, and alternative cost recovery mechanisms represent evolutionary changes that preserve the regulated monopoly structure while improving its performance and responsiveness.

The integration of distributed energy resources and new technologies will require continued evolution of utility business models and regulatory approaches to ensure that these resources contribute to rather than detract from system reliability and customer value. This integration may eventually lead to more fundamental changes in how utilities are organized and regulated.

Environmental and climate considerations will continue to drive utility investments and regulatory decisions, potentially accelerating the pace of change and increasing the importance of long-term planning and stakeholder engagement. The success of regulated utilities in adapting to these challenges will largely determine whether the regulated monopoly model remains viable or whether alternative approaches will become necessary.

Chapter 20: The ERCOT Model - A Free Market Grid in Texas

The Electric Reliability Council of Texas (ERCOT) represents one of the most ambitious and sustained experiments in competitive electricity markets in the United States, operating as an "energy-only" market that relies primarily on market forces rather than regulatory intervention to ensure adequate supply and system reliability. Since its establishment as a competitive market in 2002, ERCOT has provided valuable lessons about the benefits and challenges of market-based approaches to electricity system operation while serving as a testing ground for innovative market designs and technologies that have influenced electricity policy nationwide.

The Origins and Philosophy of ERCOT

ERCOT's origins trace back to the unique political and economic culture of Texas, which has historically favored market-based solutions over government regulation while maintaining a strong commitment to state sovereignty and independence from federal oversight. The decision to create a competitive electricity market in Texas reflected these values while responding to concerns about the efficiency and innovation of the state's regulated utility monopolies.

The Texas legislature's decision to restructure the state's electricity industry was influenced by the apparent success of deregulation in other industries and the promise that competition would deliver lower prices, improved efficiency, and greater innovation than traditional regulation could provide. Unlike other states that moved cautiously toward partial deregulation, Texas embraced a comprehensive approach that separated generation,

185

transmission, and retail sales into distinct competitive or regulated segments.

The energy-only market design adopted by ERCOT differs fundamentally from the capacity market structures used in other regions, relying entirely on energy prices to provide signals for both short-term operations and long-term investment decisions. This approach reflects a philosophical commitment to market mechanisms and a skepticism about regulatory interventions that might distort market signals or create perverse incentives.

ERCOT's geographic isolation from other interconnected power systems was both an advantage and a challenge for market development. The isolation simplified market design by eliminating the need to coordinate with neighboring regions that might have different market rules or regulatory structures, but it also meant that Texas would have limited access to external resources during emergency conditions that might threaten system reliability.

The creation of ERCOT as a nonprofit corporation governed by a board of directors with representation from market participants, consumer groups, and independent members reflected an attempt to balance different stakeholder interests while maintaining independence from political interference. This governance structure was designed to ensure that market operations would be guided by technical and economic considerations rather than political pressures.

Market Structure and Operation

ERCOT operates as both a Regional Transmission Organization (RTO) responsible for maintaining system reliability and a market operator that facilitates competitive wholesale electricity transactions. This dual role requires balancing the technical requirements of reliable system operation with the economic efficiency objectives of competitive markets, a challenge that has required continuous refinement of market rules and procedures.

The ERCOT market uses a security-constrained economic dispatch model that selects the lowest-cost combination of generation resources to meet demand while respecting transmission constraints and reliability requirements. This approach optimizes economic efficiency while maintaining the physical requirements for reliable system operation, though the balance between economic and reliability objectives can create tensions during stressed system conditions.

Real-time pricing throughout the ERCOT system provides locational price signals that reflect the marginal cost of serving load at different points in the network, including the costs of transmission congestion and losses. These price signals are designed to encourage efficient generation investment and consumption decisions while providing incentives for market participants to locate resources in ways that support rather than stress the transmission system.

The energy-only market design means that generators are compensated only for the energy they produce, rather than receiving separate payments for maintaining capacity availability as in other market designs. This approach relies on scarcity pricing during tight supply conditions to provide the revenue signals needed to encourage adequate investment in generating capacity and demand response resources.

ERCOT's ancillary services markets procure the reserves and other services needed to maintain system reliability, including regulation service to maintain frequency, spinning reserves for contingency response, and non-spinning reserves for longer-term supply security. These markets use competitive bidding to select the most cost-effective resources for providing these essential services.

Successes of the ERCOT Model

The ERCOT market has achieved significant successes that demonstrate the potential benefits of competitive electricity markets when properly designed and implemented. Competition has driven substantial efficiency improvements in both generation and retail services while encouraging innovation and investment in new technologies that might not have been adopted as quickly under traditional regulation.

Generation costs have declined significantly under competition as older, less efficient power plants have been retired or upgraded while new, more efficient facilities have been built to serve growing demand. The competitive pressure to minimize operating costs has encouraged plant owners to improve maintenance practices, optimize operating procedures, and invest in equipment upgrades that improve efficiency and reliability.

Retail competition has provided customers with choices regarding their electricity suppliers, rate structures, and service options that were not available under regulated monopoly structures. Competitive retail providers have developed innovative product offerings, including renewable energy options, demand response programs, and time-of-use rates that enable customers to manage their electricity costs while supporting system efficiency.

The ERCOT market has been particularly successful in accommodating large amounts of renewable energy, especially wind power, which has grown from virtually zero to providing over 25 percent of ERCOT's annual energy supply. The market's energy-only design and real-time pricing provide strong incentives for renewable energy development while encouraging the flexible operation of conventional resources needed to accommodate variable renewable output.

Innovation in market design and technology has flourished under the ERCOT model, with new approaches to demand response, energy storage, and distributed resource integration being tested and implemented more rapidly than might have occurred under traditional regulatory processes. The market's flexibility and responsiveness have enabled the integration of new technologies and business models that support system efficiency and reliability.

Investment in new generating capacity has generally kept pace with load growth and resource retirements, demonstrating that market signals can provide adequate incentives for resource adequacy without regulatory mandates or capacity payments. The diversity of generation technologies and ownership structures in ERCOT has increased over time, reducing concentration risks while improving system resilience.

Challenges and Criticisms

Despite its successes, the ERCOT model faces significant challenges and has been subject to substantial criticism, particularly regarding price volatility, resource adequacy, and the social distribution of market outcomes. These challenges highlight fundamental tensions between market efficiency and other policy objectives that competitive markets alone may not address adequately.

Price volatility is an inherent feature of energy-only markets, with prices ranging from near zero during periods of abundant supply to several thousand dollars per megawatt-hour during scarcity conditions. While this volatility provides important economic signals, it creates financial risks for both generators and consumers that can be difficult to manage, particularly for customers who are not sophisticated energy market participants.

Resource adequacy concerns have intensified as the ERCOT system has experienced several periods of tight supply conditions that required emergency procedures to maintain system reliability. Critics argue that the energy-only market design provides insufficient incentives for maintaining adequate reserve margins, particularly during periods of moderate market conditions that provide neither scarcity pricing nor strong capacity investment signals.

The February 2021 winter storm that caused widespread blackouts throughout Texas exposed significant vulnerabilities in the ERCOT system and market design. The extreme weather conditions caused numerous generating units to fail while demand reached record levels, creating a system emergency that required rotating outages to prevent total system collapse. The crisis highlighted questions about generator winterization requirements, market design during emergency conditions, and the adequacy of planning for extreme weather events.

Market concentration in some segments of the ERCOT market has raised concerns about the potential for the exercise of market power, particularly during periods of tight supply when a small number of generators might be able to influence prices significantly. While market monitoring and mitigation measures exist, critics argue that these protections may be insufficient during the extreme conditions that create the greatest potential for abuse.

The social distribution of market outcomes has been controversial, with critics arguing that competitive markets primarily benefit large commercial and industrial customers who can navigate complex market structures while imposing costs and risks on smaller customers who have limited market power or sophistication. Low-income customers may be particularly vulnerable to price volatility and marketing practices that are not always transparent or fair.

The February 2021 Crisis - A Defining Moment

The February 2021 winter storm represents the most significant challenge that the ERCOT system and market design have faced, providing crucial lessons about the vulnerabilities of energy-only markets during extreme conditions while highlighting the importance of adequate planning and preparation for unusual events. The crisis began when an unprecedented winter storm brought record cold temperatures to Texas, creating enormous demand for heating while simultaneously causing widespread failures of generating equipment that was not designed for such extreme conditions.

The failure of generating units during the crisis was caused by multiple factors, including inadequate winterization of equipment, fuel supply disruptions, and the decision by some plant operators to take units offline for maintenance during winter months when demand was typically lower. Natural gas-fired power plants were particularly affected as frozen equipment and pipeline constraints limited fuel availability precisely when demand was highest.

Wind power generation also declined during the storm, though by a smaller amount than anticipated and much less than the outages experienced by conventional generation. The reduction in wind output became a focal point for political criticism of renewable energy, though subsequent analysis revealed that failures of conventional generation were the primary cause of the supply shortage.

The ERCOT system operator was forced to implement rotating outages that were intended to last for short periods but extended for days in some areas as the supply shortage persisted and additional generators failed. These outages caused widespread hardship, economic losses, and in some cases contributed to deaths from carbon monoxide poisoning and hypothermia as people sought alternative heating sources.

Wholesale electricity prices reached the administrative cap of $9,000 per megawatt-hour and remained there for several days, creating billions of dollars in costs that were ultimately allocated among market participants according to complex market rules. The price spikes enriched some market participants while threatening the financial viability of others, highlighting questions about risk allocation and market design during emergency conditions.

Regulatory and Market Responses

The February 2021 crisis prompted extensive regulatory and legislative responses designed to address the vulnerabilities exposed by the winter storm while preserving the competitive market structure that Texas policymakers continue to support. These responses have focused on improving system resilience, enhancing emergency preparedness, and modifying market rules to better align private incentives with public objectives.

Generator winterization requirements were significantly enhanced, with mandatory weatherization standards and inspections designed to ensure that power plants can operate reliably during extreme weather conditions. These requirements impose additional costs on generators but are intended to reduce the risk of widespread equipment failures during future extreme weather events.

Market design modifications have been implemented to improve incentives for resource adequacy and system reliability, including changes to administrative price caps, reserve margin requirements, and emergency pricing procedures. These modifications attempt to address some of the resource adequacy concerns that contributed to the February 2021 crisis while preserving the energy-only market structure.

Enhanced coordination with natural gas infrastructure has been implemented to improve fuel supply reliability during extreme weather events. This coordination includes information sharing agreements, joint planning processes, and priority service arrangements designed to ensure that critical power plants can access natural gas supplies during emergencies.

Emergency preparedness and response procedures have been improved based on lessons learned from the February 2021 crisis, including better communication protocols, enhanced load forecasting capabilities, and improved coordination among different agencies and jurisdictions responsible for emergency management.

Renewable Energy and Grid Integration

ERCOT has become a leading laboratory for renewable energy integration, demonstrating both the opportunities and challenges associated with high levels of variable renewable generation. Wind power has grown dramatically in ERCOT, benefiting from excellent wind resources, competitive market structures, and federal tax incentives that have made wind power one of the lowest-cost generation sources in the region.

The integration of large amounts of wind power has required modifications to system operations and market designs to accommodate the variable and sometimes unpredictable nature of wind generation. ERCOT has developed sophisticated forecasting capabilities, flexible market structures, and operational procedures that enable the reliable integration of renewable resources while maintaining system stability.

Solar power development has accelerated more recently, with declining costs and improved technology making solar competitive with other generation sources in many parts of Texas. The growth of solar power creates both opportunities and challenges that differ from those associated with wind power, requiring continued evolution of system operations and market designs.

Energy storage deployment is increasing rapidly in ERCOT as battery costs decline and market rules are modified to enable storage resources to participate effectively in energy and ancillary services markets. Storage resources can provide valuable services for renewable energy integration while offering new approaches to managing system reliability and market efficiency.

The combination of renewable energy growth and energy storage deployment is beginning to transform the ERCOT system in fundamental ways, creating new patterns of generation and consumption while requiring continued adaptation of market rules and operational procedures. These changes represent both opportunities for continued cost reductions and challenges for maintaining system reliability.

Lessons for Other Jurisdictions

The ERCOT experience provides valuable lessons for other jurisdictions considering competitive electricity markets or modifications to existing market designs. The successes of ERCOT demonstrate the potential benefits of competitive markets, while its challenges highlight the importance of careful market design and ongoing adaptation to changing conditions.

The energy-only market design can work effectively under normal conditions, providing strong incentives for efficiency and innovation while accommodating new technologies and business models. However, the design may require modifications or supplementary mechanisms to address resource adequacy concerns and ensure adequate system resilience during extreme conditions.

Market isolation can simplify market design and operation but creates vulnerabilities during emergency conditions when external resources might otherwise be available to help maintain system reliability. The costs and benefits of market integration versus isolation must be carefully weighed based on each jurisdiction's specific circumstances and policy objectives.

Political and regulatory commitment to market principles is essential for competitive markets to function effectively, but this commitment must be balanced against other policy objectives including affordability, reliability, and environmental protection. The tension between market efficiency and other objectives requires ongoing attention and may necessitate market design modifications over time.

The ERCOT model continues to evolve in response to changing conditions and lessons learned from both successes and failures. This ongoing evolution demonstrates both the flexibility of competitive markets and the need for continuous monitoring, evaluation, and adaptation to ensure that markets serve the public interest while achieving their efficiency and innovation objectives.

Chapter 21: Renewable Energy Integration - A New Era for the Grid

The integration of renewable energy sources into the American electrical grid represents one of the most significant transformations in the industry's history, fundamentally altering how electricity is generated, transmitted, and consumed while creating unprecedented technical, economic, and policy challenges. The rapid growth of wind, solar, and other renewable technologies has been driven by dramatic cost reductions, environmental concerns, and supportive government policies that have made renewable energy competitive with conventional generation sources in many markets. This transformation is reshaping every aspect of grid operations while requiring new approaches to planning, investment, and regulation that can accommodate the unique characteristics of renewable energy resources.

The Renewable Energy Revolution

The growth of renewable energy in the United States has exceeded the most optimistic projections made just two decades ago, transforming from a niche sector serving specialized markets to a mainstream industry that competes directly with conventional generation technologies. Wind power capacity has grown from less than 3,000 megawatts in 2000 to over 130,000 megawatts today, while solar capacity has expanded even more dramatically from virtually zero to over 100,000 megawatts during the same period.

This remarkable growth has been driven primarily by dramatic cost reductions that have made renewable technologies competitive with fossil fuel alternatives in many markets. Wind power costs have declined by approximately 70 percent since 2010, while solar photovoltaic costs have fallen by over 80 percent during the same period. These cost reductions reflect improvements in manufacturing, technology advances, economies of scale, and learning curve effects that continue to drive further cost improvements.

Government policies have played a crucial role in supporting renewable energy development through production tax credits, investment tax credits, renewable portfolio standards, and other incentives that have helped emerging technologies achieve commercial viability. Federal tax incentives have been particularly important for wind and solar development, providing financial support that has enabled these technologies to compete with conventional generation while achieving the scale necessary for continued cost reductions.

State renewable portfolio standards have created sustained demand for renewable energy by requiring utilities to procure specified percentages of their electricity from renewable sources. These mandates have provided long-term market certainty that has encouraged investment in renewable energy projects while driving technological innovation and cost reductions throughout the industry.

Environmental concerns about climate change and air pollution have created additional drivers for renewable energy development as policymakers and consumers seek cleaner alternatives to fossil fuel generation. The Biden administration has established a goal of achieving 100 percent carbon-free electricity by 2035, creating further momentum for renewable energy deployment while requiring unprecedented levels of investment in both generation and transmission infrastructure.

Technical Challenges of Integration

The integration of large amounts of renewable energy into the electrical grid creates unprecedented technical challenges that require new approaches to system planning, operations, and control. Unlike conventional power plants that can be dispatched on demand to match supply with demand, renewable energy sources are variable and partially unpredictable, generating electricity only when wind blows or sun shines while sometimes producing more power than the system can immediately use.

The variability of renewable generation creates new demands for flexibility in system operations, requiring resources that can quickly increase or decrease their output to compensate for changes in renewable generation. This flexibility can come from conventional power plants operating at part load, energy storage systems, demand response programs, or transmission interconnections that enable sharing of resources across broader geographic areas.

Grid stability becomes more challenging with high levels of renewable penetration because many renewable technologies do not provide the same grid support services that conventional synchronous generators have historically provided. Wind and solar generators typically connect to the grid through power electronic interfaces that do not naturally provide inertia, voltage support, or other ancillary services that are essential for reliable grid operations.

Transmission planning must account for the geographic distribution of renewable resources, which are often located in areas with limited existing transmission infrastructure that are far from population centers where electricity is consumed. Wind resources are generally best in the Great Plains and offshore areas, while solar resources are most abundant in the Southwest and Southeast, requiring new transmission lines to connect these resources to load centers.

Distribution system impacts become significant as rooftop solar and other distributed renewable resources change power flows on local distribution networks that were designed for one-way power flow from central substations to customers. High penetrations of distributed solar can create voltage regulation problems, equipment overloading, and safety concerns that require system modifications and new operational procedures.

Forecasting renewable generation output has become essential for system operations but remains challenging due to the inherent variability and unpredictability of weather conditions. Improved weather forecasting, statistical modeling, and machine learning techniques have enhanced renewable forecasting accuracy, but uncertainty remains a fundamental characteristic that system operators must manage.

Grid Modernization Requirements

The integration of renewable energy requires comprehensive grid modernization that goes far beyond simply connecting new generation sources to existing transmission and distribution networks. Modern grids must be more flexible, intelligent, and resilient than traditional systems while accommodating two-way power flows, variable generation sources, and active customer participation in grid operations.

Smart grid technologies are essential for managing the complexity of modern power systems with high levels of renewable penetration. Advanced sensors, communication systems, and control technologies enable real-time monitoring and control of grid conditions while providing the visibility and responsiveness needed to accommodate variable renewable generation.

Energy storage systems are increasingly recognized as crucial for renewable energy integration, providing services that include energy shifting from periods of high renewable output to periods of high demand, frequency regulation to maintain grid stability, and backup power during renewable generation lulls. Battery storage costs have declined dramatically in recent years, making storage economically viable for an increasing range of applications.

Transmission system expansion is essential for accessing the best renewable resources and sharing renewable generation across broader geographic areas to smooth out variability. Studies suggest that over a million miles of new transmission will need to be built by 2050 to support renewable energy goals, representing an unprecedented expansion of the transmission system.

Distribution system upgrades are required to accommodate distributed renewable resources while maintaining safety and reliability standards. These upgrades include advanced inverters, voltage regulation equipment, protection system modifications, and communication networks that enable coordination between distributed resources and system operators.

Grid-forming inverters and other advanced technologies are being developed to enable renewable resources to provide the grid support services traditionally provided by conventional generators. These technologies could eventually enable power systems to operate reliably with very high levels of renewable penetration while maintaining the stability and resilience that customers expect.

Economic Impacts and Market Evolution

The rapid growth of renewable energy is transforming electricity markets in fundamental ways that affect pricing patterns, investment decisions, and competitive dynamics throughout the industry. The near-zero marginal costs of wind and solar generation are displacing higher-cost conventional resources while creating new patterns of wholesale electricity prices that reflect the timing and variability of renewable generation.

Merit order effects occur when renewable generation displaces conventional generation in economic dispatch, reducing wholesale electricity prices during periods of high renewable output. These effects can significantly reduce electricity market revenues for all generators, creating challenges for conventional power plants that may still be needed for reliability purposes during periods of low renewable generation.

Capacity value of renewable resources differs from their energy value because renewable output may not be well-correlated with peak demand periods when system reliability is most at risk. Wind generation often peaks at night when demand is low, while solar generation peaks during midday when demand may be moderate rather than at evening peak hours when demand is typically highest.

Market design evolution is necessary to accommodate the unique characteristics of renewable resources while maintaining incentives for system reliability and long-term resource adequacy. This evolution includes modifications to capacity markets, ancillary service markets, and transmission planning processes that better reflect the contributions and requirements of renewable resources.

Investment patterns are shifting as renewable energy attracts increasing amounts of capital while conventional generation faces declining capacity factors and revenues. This transition requires careful management to ensure that adequate conventional resources remain available during the transition period while avoiding excessive costs from maintaining unneeded capacity.

Regional differences in renewable resource quality and policy environments are creating new patterns of interstate electricity trade as areas with excellent renewable resources become net exporters while areas with limited renewable potential become net importers. These changing trade patterns require enhanced transmission infrastructure and coordination mechanisms.

Policy and Regulatory Adaptation

The integration of renewable energy requires comprehensive policy and regulatory reforms that address barriers to deployment while ensuring that renewable resources contribute appropriately to system reliability and costs. Traditional regulatory frameworks were designed for conventional resources and may not provide appropriate incentives or requirements for renewable resources with different operating characteristics.

Interconnection procedures and standards must be updated to streamline the process of connecting renewable resources to the grid while ensuring that these resources meet technical requirements for safe and reliable operation. Lengthy interconnection queues and costly system upgrade requirements have become significant barriers to renewable development in many regions.

Transmission planning and cost allocation mechanisms must evolve to accommodate the geographic distribution of renewable resources and the benefits they provide to multiple states and regions. Traditional planning processes that focus on local reliability needs may not adequately consider the regional benefits of transmission lines that access remote renewable resources.

Net metering and distributed resource compensation policies must balance the interests of customers who invest in distributed renewable resources with the interests of other customers who may bear the costs of maintaining the grid infrastructure that supports these resources. These policies have become increasingly controversial as distributed renewable penetration grows.

Market design reforms are necessary to ensure that renewable resources face appropriate incentives to contribute to system reliability while receiving fair compensation for the services they provide. These reforms may include modifications to capacity markets, ancillary service markets, and transmission planning processes.

Environmental and permitting policies must be streamlined to facilitate renewable energy deployment while protecting environmental resources and community interests. The permitting process for large renewable projects and transmission lines can be lengthy and complex, creating barriers to the rapid deployment needed to meet climate goals.

Storage Technologies and Grid Services

Energy storage technologies have emerged as essential complements to renewable energy, providing services that enable higher levels of renewable penetration while improving overall system flexibility and reliability. Battery storage costs have declined by over 80 percent since 2010, making storage economically viable for an increasing range of applications from utility-scale grid services to residential backup power.

Grid-scale battery storage is being deployed rapidly to provide services including frequency regulation, spinning reserves, peak shaving, and renewable energy shifting. These applications take advantage of batteries' ability to respond quickly and precisely to control signals while providing multiple services from the same asset.

Pumped hydroelectric storage represents the largest existing storage resource and continues to play an important role in renewable integration, particularly for long-duration energy storage that can shift renewable energy across many hours or even days. However, suitable sites for new pumped storage facilities are limited, constraining expansion of this technology.

Emerging storage technologies including compressed air energy storage, flow batteries, and hydrogen production offer potential solutions for long-duration storage needs that batteries may not be able to meet economically. These technologies are still in early stages of commercial deployment but could become important for achieving very high levels of renewable penetration.

Vehicle-to-grid technologies could eventually enable electric vehicle batteries to provide grid services when vehicles are parked and connected to the grid. This application could provide enormous storage capacity as electric vehicle adoption grows, though technical and regulatory barriers must be overcome to realize this potential.

Storage market participation rules are evolving to enable storage resources to provide multiple services while being appropriately compensated for the value they provide. These rule changes are necessary to ensure that storage investments are economically viable while contributing effectively to system operations.

Future Outlook and Challenges

The continued growth of renewable energy appears likely given ongoing cost reductions, environmental pressures, and policy support, but achieving very high levels of renewable penetration will require addressing increasingly challenging technical and economic issues. Current renewable penetration levels of 20-30 percent in some regions are already creating significant operational challenges that will intensify as penetration increases.

System flexibility will become increasingly valuable and necessary as renewable penetration grows, requiring investments in storage, demand response, transmission, and flexible conventional generation that can accommodate large and rapid changes in renewable output. The costs and benefits of these flexibility investments must be carefully managed to maintain system reliability while avoiding excessive costs.

Seasonal and long-duration storage needs will become more significant as renewable energy provides larger shares of annual electricity generation. Current battery technologies are well-suited for daily energy shifting and short-term grid services but may not be economical for storing renewable energy across weeks or months.

Grid resilience and security concerns may increase with high levels of renewable penetration as the system becomes more dependent on weather conditions and electronic systems that could be vulnerable to cyber attacks or equipment failures. These concerns require careful attention to system design and emergency preparedness.

Social and environmental justice considerations must be addressed as the transition to renewable energy affects different communities and regions differently. Areas dependent on fossil fuel industries may face economic challenges, while communities near renewable projects may experience both benefits and impacts that require careful management.

The renewable energy transformation represents both an unprecedented opportunity to create a cleaner, more sustainable electricity system and a complex challenge that requires careful planning, substantial investment, and continued innovation. Success will require coordinated efforts among policymakers, regulators, utilities, technology developers, and consumers to ensure that the transition benefits society while maintaining the reliable, affordable electricity service that modern society requires.

Chapter 22: Smart Grid Technologies - Modernizing the Infrastructure

The evolution of the American electrical grid from a collection of mechanical switches and analog controls to a sophisticated digital network represents one of the most significant technological transformations in the industry's history. Smart grid technologies promise to revolutionize how electricity is generated, transmitted, distributed, and consumed by leveraging advanced sensors, communications systems, data analytics, and automated controls to create a more efficient, reliable, and responsive electrical system. However, this digital transformation also introduces new vulnerabilities and challenges that must be carefully managed to ensure that modernization efforts enhance rather than compromise grid security and reliability.

Defining the Smart Grid

The concept of a "smart grid" encompasses a broad range of technologies and capabilities that transform the traditional electrical grid from a passive, one-way system into an active, two-way network that can respond dynamically to changing conditions and customer needs. At its core, a smart grid uses digital communication technology to monitor and manage the flow of electricity efficiently, enabling real-time adjustments and optimizations that ensure stable and reliable power supply.

Unlike traditional grids that operate largely passively, smart grids are dynamic and interactive, allowing for bidirectional communication between utilities and consumers while providing unprecedented visibility into system operations at every level from individual customer meters to major transmission facilities.

This enhanced visibility enables system operators to identify and respond to problems more quickly while optimizing system performance in ways that were previously impossible.

The smart grid represents a convergence of electrical engineering, information technology, and telecommunications that creates new capabilities for system monitoring, control, and optimization. Advanced metering infrastructure forms the foundation of smart grid deployments, providing detailed real-time data about electricity consumption patterns while enabling new services and rate structures that can better align customer behavior with system needs.

Communication networks are essential for smart grid operations, connecting millions of devices and sensors throughout the electrical system while enabling the rapid exchange of information needed for real-time control and optimization. These networks must be reliable, secure, and capable of handling the enormous volumes of data generated by smart grid systems while maintaining the low latency required for critical control applications.

Data analytics and artificial intelligence capabilities are increasingly important for extracting value from the vast amounts of information collected by smart grid systems. These technologies can identify patterns, predict equipment failures, optimize system operations, and enable new services that improve both system efficiency and customer satisfaction.

Core Components and Technologies

Advanced metering infrastructure represents the most visible component of smart grid deployments, replacing traditional mechanical meters with digital devices that can measure and communicate detailed information about electricity usage patterns. Smart meters provide customers with near real-time information about their electricity consumption while enabling utilities to implement time-of-use rates, demand response programs, and other advanced services that can improve system efficiency.

The deployment of smart meters has been extensive throughout the United States, with over 100 million smart meters installed as of 2024, representing approximately 70 percent of all electricity customers. These deployments have enabled new capabilities including remote meter reading, outage detection, and customer engagement programs while providing the data foundation for more advanced smart grid applications.

Sensors and monitoring equipment throughout the transmission and distribution systems provide real-time visibility into system conditions that enable more precise control and faster response to changing conditions. These devices can monitor everything from transformer loading and power quality to weather conditions and equipment health, providing system operators with unprecedented situational awareness.

Distribution automation systems use advanced sensors, communications, and control devices to enable automatic responses to system disturbances such as equipment failures or storm damage. These systems can isolate problems and restore service to unaffected areas within minutes rather than hours, significantly improving system reliability while reducing the duration and scope of outages.

Supervisory control and data acquisition (SCADA) systems have been enhanced with modern communication technologies and cybersecurity measures to provide more robust and secure control of critical grid infrastructure. Modern SCADA systems can integrate information from thousands of remote devices while providing operators with intuitive interfaces for monitoring and controlling system operations.

Energy management systems integrate data from throughout the electrical system to optimize generation dispatch, manage transmission constraints, and coordinate system operations in real-time. These systems use sophisticated algorithms to balance supply and demand while minimizing costs and maintaining reliability standards.

Demand response technologies enable customers to participate actively in grid operations by adjusting their electricity consumption in response to system conditions or price signals. These technologies can range from simple programmable thermostats to sophisticated energy management systems that can automatically modify industrial processes based on grid conditions.

Benefits of Smart Grid Implementation

The deployment of smart grid technologies offers numerous benefits that can improve system efficiency, reliability, customer service, and environmental performance while enabling new business models and services that were not possible with traditional grid infrastructure. These benefits justify the substantial investments required for smart grid modernization while providing value to both utilities and their customers.

Improved system efficiency results from better visibility and control capabilities that enable more precise matching of supply and demand while reducing losses throughout the transmission and distribution systems. Smart grid technologies can optimize power flows, reduce peak demand through demand response programs, and enable more efficient integration of distributed energy resources such as rooftop solar and energy storage systems.

Enhanced reliability comes from faster detection and response to system problems, automated restoration capabilities, and improved situational awareness that enables proactive maintenance and system optimization. Smart grid systems can isolate problems more quickly, reroute power around damaged equipment, and restore service to customers faster than traditional systems while reducing the frequency and duration of outages.

Better customer service is enabled by improved communication capabilities, more detailed usage information, and new service offerings that give customers greater control over their electricity costs and consumption patterns. Smart grid technologies enable time-of-use rates, prepaid service options, and energy efficiency programs that can help customers manage their electricity bills while supporting overall system efficiency.

Environmental benefits result from improved system efficiency, better integration of renewable energy sources, and reduced need for backup generation during peak demand periods. Smart grid technologies can accommodate higher levels of variable renewable generation while enabling demand response programs that reduce the need for fossil fuel-fired peaking power plants.

Grid modernization also enables new business models and services including distributed energy resource integration, electric vehicle charging management, and enhanced energy storage utilization that can create value for both utilities and customers while supporting broader energy policy objectives such as decarbonization and energy security.

Economic benefits include reduced operational costs, deferred infrastructure investments, and improved asset utilization that can help keep electricity rates affordable while maintaining high levels of service quality. Smart grid investments can pay for themselves through operational savings while providing capabilities that enable future innovations and services.

Cybersecurity Challenges in the Digital Age

The digitization of the electrical grid has created unprecedented cybersecurity challenges that represent one of the most significant risks facing the modern power system. As utilities have upgraded systems and exposed operational technology to internet connectivity to address the challenges of managing geographically complex systems, the attack surface for malicious actors has expanded dramatically.

The increasing connectivity of smart grid systems introduces new vulnerabilities that did not exist in traditional electrical systems, which were largely isolated from external networks and relied on physical security rather than cybersecurity for protection. Modern smart grids depend on complex networks of sensors, controllers, and communication systems that create multiple potential entry points for cyber attackers while connecting critical infrastructure to potentially vulnerable external networks.

Smart grid communication networks are particularly vulnerable to various types of cyberattacks including denial-of-service attacks that can disrupt communications, replay attacks that can manipulate control signals, time synchronization attacks that can disrupt coordination between different parts of the system, false data injection attacks that can corrupt system information, and malware that can compromise the operation of critical control systems.

The 2015 cyberattack on Ukraine's power grid demonstrated the potential for sophisticated adversaries to cause widespread blackouts by infiltrating human-machine interfaces and manipulating control systems. This attack, which caused a 10-hour blackout affecting over 100,000 people, highlighted the vulnerability of modern power systems to coordinated cyber attacks while demonstrating the potential for cyber warfare to target critical infrastructure.

Recent cybersecurity incidents in 2025 have further highlighted these vulnerabilities, including ransomware attacks on energy contractors like ENGlobal Corporation that disrupted operations for weeks, breaches of communication networks critical to energy infrastructure by Chinese hackers, and data exposures at utility companies that compromised customer information.

The challenge is compounded by the fact that utilities may not own or directly control many of the technologies and systems being connected to the grid, as distributed energy resources, smart home devices, and other customer-owned equipment create new attack vectors that are outside traditional utility security frameworks. This expanded threat landscape requires new approaches to cybersecurity that extend beyond traditional utility boundaries.

Government agencies including the Cybersecurity and Infrastructure Security Agency (CISA) have noted that "foreign adversaries continue to explore ways to access critical infrastructure," while international organizations have urged the adoption of AI-based anomaly detection systems for industrial control systems. The development of new attack techniques specifically targeting energy sector infrastructure has led to expanded threat intelligence frameworks and defensive strategies.

Cybersecurity Solutions and Defensive Measures

Addressing smart grid cybersecurity challenges requires comprehensive defensive strategies that combine technology solutions, operational procedures, and organizational changes to create multiple layers of protection against increasingly sophisticated threats. These defensive measures must evolve continuously as attack techniques become more advanced while balancing security requirements with operational efficiency and cost considerations.

Network segmentation represents a fundamental defensive strategy that involves separating information technology (IT) networks from operational technology (OT) systems to prevent attackers from moving laterally through interconnected systems. This approach limits the potential impact of successful attacks while providing additional barriers that attackers must overcome to access critical control systems.

Zero Trust security models are being adopted by utilities to verify every device and user attempting to access grid systems regardless of their location or previous authentication status. This approach assumes that threats may already exist within network perimeters and requires continuous verification and monitoring of all access attempts and system interactions.

Advanced threat detection systems use artificial intelligence and machine learning to identify unusual patterns of behavior that may indicate cyber attacks or system compromises. These systems can analyze vast amounts of data from throughout the smart grid infrastructure to detect subtle signs of malicious activity that human operators might miss while providing rapid alerts that enable quick response to potential threats.

Digital twin technologies create virtual replicas of physical grid systems that can be used to test security measures, simulate the impact of potential attacks, and develop response strategies without affecting actual grid operations. These simulations enable utilities to understand vulnerabilities and test defensive measures in a safe environment while improving their preparedness for real-world attacks.

Encryption and authentication technologies protect communications between smart grid devices and control systems while ensuring that only authorized users and devices can access critical systems. Modern encryption standards and multi-factor authentication systems provide strong protection against unauthorized access while maintaining the performance required for real-time grid operations.

Regular penetration testing and security assessments help utilities identify vulnerabilities in their systems before attackers can exploit them. These assessments should include both IT and OT systems while considering the unique characteristics of smart grid infrastructure and the potential impact of successful attacks on public safety and grid reliability.

Supply chain security has become increasingly important as smart grid deployments rely on equipment and software from numerous vendors that may have different security standards and practices. Utilities must carefully evaluate the cybersecurity capabilities of their suppliers while implementing security requirements and ongoing monitoring to ensure that third-party systems do not introduce vulnerabilities.

Advanced Metering Infrastructure and Customer Engagement

Advanced metering infrastructure forms the cornerstone of most smart grid deployments, providing the data foundation and customer interface that enables many advanced smart grid applications and services. The widespread deployment of smart meters has transformed how utilities interact with customers while creating new opportunities for energy efficiency, demand response, and customer engagement programs that can benefit both utilities and their customers.

Smart meters provide customers with detailed, near real-time information about their electricity consumption that was never available with traditional mechanical meters. This information can help customers understand how their behavior affects their electricity bills while identifying opportunities for energy savings through behavioral changes or equipment upgrades. The availability of detailed usage data has enabled new rate structures and programs that provide customers with greater control over their electricity costs.

Time-of-use rates enabled by smart meters can provide customers with price signals that reflect the actual cost of providing electricity at different times of day, encouraging consumption during periods when electricity is less expensive while reducing demand during peak periods when system costs are highest. These rate structures can provide significant savings for customers who can shift their electricity usage to off-peak periods while helping utilities manage peak demand more effectively.

Demand response programs use smart meter data and communication capabilities to enable customers to reduce their electricity consumption during periods of high system demand or supply shortage. These programs can range from simple alerts that inform customers about high-cost periods to automated systems that can adjust thermostats, water heaters, and other equipment in response to system conditions or price signals.

Prepaid electricity service options enabled by smart meters allow customers to pay for electricity in advance rather than receiving monthly bills, providing better budget control for some customers while reducing bad debt risk for utilities. These programs can be particularly valuable for customers with limited credit or those who prefer to manage their electricity costs through prepayment rather than monthly billing.

Home energy management systems can integrate with smart meters and utility communication systems to provide customers with sophisticated tools for monitoring and controlling their electricity consumption. These systems can automatically adjust heating, cooling, and other systems based on time-of-use rates, renewable energy availability, or customer preferences while providing detailed feedback about energy usage patterns.

Customer engagement programs use smart meter data to provide personalized energy efficiency recommendations, usage comparisons with similar households, and other information that can help customers reduce their electricity consumption and costs. These programs have proven effective at encouraging energy conservation while improving customer satisfaction with utility services.

Integration with Renewable Energy and Storage

Smart grid technologies are essential for integrating high levels of renewable energy and energy storage into the electrical system, providing the visibility, control, and coordination capabilities needed to manage variable generation sources while maintaining system reliability and stability. The growth of distributed renewable resources has created new challenges that require sophisticated monitoring and control systems to ensure reliable grid operations.

Variable renewable generation from wind and solar sources requires advanced forecasting and control systems that can predict output levels and coordinate with other generation sources to maintain the balance between supply and demand. Smart grid systems provide the real-time data and communication capabilities needed for these coordination functions while enabling more precise control of conventional generation resources that must compensate for renewable variability.

Distributed energy resource management systems use smart grid communications and control capabilities to coordinate the operation of rooftop solar, energy storage, electric vehicles, and other customer-owned resources that can affect grid operations. These systems can optimize the output and consumption of distributed resources to support rather than stress the grid while maximizing the value these resources provide to both customers and the overall system.

Energy storage integration requires sophisticated control systems that can coordinate battery charging and discharging with renewable generation output, system demand, and market conditions to maximize the value of storage investments while providing grid services such as frequency regulation and peak load management. Smart grid systems provide the communication and control infrastructure needed for these complex optimization functions.

Grid-scale renewable projects benefit from smart grid monitoring and control capabilities that enable better integration with transmission systems while providing system operators with the information needed to manage power flows and maintain system stability. Advanced power electronics and control systems enable renewable generators to provide grid support services that were traditionally provided only by conventional power plants.

Microgrids and distributed energy systems use smart grid technologies to operate semi-independently from the main grid while providing enhanced reliability and resilience for critical facilities or communities. These systems require sophisticated control and protection systems that can seamlessly transition between grid-connected and islanded operation while maintaining power quality and safety standards.

Economic Considerations and Investment Challenges

The deployment of smart grid technologies requires substantial capital investments that must be justified through demonstrable benefits to customers and society while considering the long-term nature of electrical infrastructure and the uncertainty surrounding future technological developments. Utilities and regulators must carefully evaluate the costs and benefits of smart grid investments while ensuring that these investments serve the public interest.

Smart meter deployments have required billions of dollars in investments that have been controversial in some jurisdictions due to questions about cost-effectiveness, customer benefits, and the allocation of costs among different customer classes. While smart meters enable numerous advanced applications and services, the direct benefits to customers may not always be immediately apparent or sufficient to justify the full cost of deployment.

Grid modernization investments beyond basic smart meters can be even more challenging to justify economically, as the benefits of improved reliability, enhanced cybersecurity, and renewable energy integration may be difficult to quantify in traditional cost-benefit analyses. These investments often provide system-wide benefits that are not easily attributed to specific customers or services while requiring upfront costs that may not be recovered for many years.

Cost recovery mechanisms for smart grid investments vary among jurisdictions but typically involve some combination of rate base inclusion for capital investments, performance incentives tied to specific outcomes, and shared savings mechanisms that allow utilities to retain a portion of the cost savings generated by smart grid technologies. These mechanisms must balance utility cost recovery needs with customer protection while providing incentives for efficient and effective smart grid deployment.

Stranded asset risks can arise from smart grid investments as rapidly evolving technology may make some investments obsolete before they reach the end of their useful lives. Utilities must carefully consider the pace of technological change and the potential for future developments when making long-term infrastructure investments while maintaining flexibility to adapt to changing conditions.

Competitive pressures from new technologies and business models enabled by smart grid capabilities can affect utility revenues and business models in ways that must be considered in investment planning and regulatory processes. The ability of customers to generate their own electricity, participate in demand response programs, and manage their consumption through smart grid technologies can reduce utility sales while requiring ongoing investments in grid infrastructure.

Looking Forward: The Future of Smart Grid Technology

The continued evolution of smart grid technologies promises to further transform the electrical industry while creating new opportunities and challenges that will require ongoing adaptation by utilities, regulators, and customers. Emerging technologies including artificial intelligence, advanced analytics, blockchain, and edge computing are likely to enable new capabilities and applications that build on current smart grid foundations.

Artificial intelligence and machine learning applications will become increasingly sophisticated in their ability to optimize grid operations, predict equipment failures, and enable autonomous responses to changing conditions. These technologies can process vast amounts of data from smart grid sensors and systems to identify patterns and optimize operations in ways that human operators could never achieve while providing predictive capabilities that enable proactive rather than reactive system management.

Edge computing technologies will enable more processing and decision-making to occur at the grid edge rather than in centralized control centers, reducing communication latency while improving system resilience and enabling faster responses to local conditions. These capabilities will be particularly important for managing high levels of distributed energy resources while maintaining system stability and reliability.

Blockchain and distributed ledger technologies may enable new business models and market mechanisms that allow peer-to-peer energy trading, automated contract execution, and enhanced cybersecurity through decentralized verification and authentication systems. While these applications are still in early development stages, they could fundamentally change how electricity markets operate and how customers interact with the grid.

The integration of smart grid technologies with other infrastructure systems including transportation, water, telecommunications, and buildings will create new opportunities for optimization and efficiency while requiring enhanced coordination and planning across multiple sectors. The electrification of transportation through electric vehicles represents one of the most significant opportunities for smart grid technologies to create value through managed charging and vehicle-to-grid services.

As smart grid technologies continue to evolve and mature, their success will depend on addressing ongoing challenges including cybersecurity risks, customer acceptance, regulatory adaptation, and economic viability while delivering demonstrable benefits that justify the substantial investments required for widespread deployment.

Chapter 23: Challenges and Opportunities - The Future of the US Electrical Grid

As the American electrical grid approaches the end of its second century of service, it faces an unprecedented convergence of challenges and opportunities that will determine its evolution for decades to come. The industry must simultaneously address aging infrastructure, climate change impacts, cybersecurity threats, and changing customer expectations while integrating new technologies and business models that promise to transform how electricity is generated, delivered, and consumed. The decisions made in the coming years will shape an electrical system that must be more reliable, sustainable, affordable, and secure than ever before while serving a society that is increasingly dependent on electricity for economic activity, public safety, and quality of life.

Infrastructure Aging and Replacement Needs

The American electrical grid's aging infrastructure represents one of the most pressing challenges facing the industry, with the majority of transmission lines, distribution equipment, and generating facilities approaching or exceeding their original design lives while serving a society that has become far more dependent on electricity than when most of the existing infrastructure was built. The scale of replacement needs is staggering, requiring unprecedented levels of investment while maintaining affordable electricity service and reliable operations during the transition period.

Transmission infrastructure built primarily in the 1960s and 1970s is now approaching the end of its 50-80 year design life, with over 70 percent of transmission lines more than 25 years old and many critical facilities over 40 years old. This aging infrastructure faces increasing maintenance costs, higher failure rates, and reduced capacity relative to modern equipment while being asked to carry more power over longer distances to serve growing demand and integrate remote renewable energy resources.

Distribution systems face similar challenges with transformers, switches, cables, and other equipment that was installed decades ago now requiring frequent maintenance and replacement while accommodating new demands from electric vehicles, distributed energy resources, and digital devices that have very different characteristics than the residential and commercial loads these systems were designed to serve. The cost of replacing distribution infrastructure is enormous, with estimates suggesting that hundreds of billions of dollars will be needed over the next two decades.

Generating facilities present complex replacement decisions as aging coal and nuclear plants reach retirement age while facing competition from lower-cost natural gas and renewable energy sources. Many of these facilities represent significant investments that have not been fully depreciated, creating stranded asset concerns while their retirement removes dispatchable capacity that may be needed for system reliability even as they become economically uncompetitive.

The challenge of infrastructure replacement is compounded by supply chain constraints, skilled worker shortages, and regulatory processes that can delay critical projects for years while costs continue to escalate. The electrical industry must develop new approaches to infrastructure planning, financing, and construction that can accelerate necessary replacements while managing costs and minimizing service disruptions.

Climate change is accelerating infrastructure aging and creating additional replacement needs as extreme weather events cause damage that would not have occurred under historical climate conditions. Heat waves stress transformers and transmission lines, severe storms cause widespread outages, wildfires threaten transmission corridors, and flooding can damage underground equipment while rising sea levels threaten coastal facilities.

Environmental and Climate Challenges

Climate change represents both one of the most significant challenges facing the electrical grid and one of its greatest opportunities to contribute to solutions through decarbonization and improved resilience. The electrical industry accounts for approximately 25 percent of U.S. greenhouse gas emissions, making it a critical sector for climate action while the grid infrastructure itself faces increasing risks from changing weather patterns and extreme events.

Decarbonization requirements are driving rapid changes throughout the electrical system as policies at federal, state, and local levels mandate reductions in carbon emissions from electricity generation. President Biden's goal of achieving 100 percent carbon-free electricity by 2035 would require unprecedented changes to the generation mix while maintaining reliability and affordability in a system that currently depends on fossil fuels for over 60 percent of electricity generation.

The transition to clean energy sources requires massive investments in renewable generation, energy storage, transmission infrastructure, and grid modernization technologies while potentially stranding existing fossil fuel investments that may become uneconomical before reaching the end of their useful lives. Managing this transition while maintaining system reliability and customer affordability presents complex technical and economic challenges.

Extreme weather events are becoming more frequent and severe, causing billions of dollars in damages to electrical infrastructure while creating widespread outages that can last for days or weeks. Hurricane damage to transmission and distribution systems, wildfire risks to transmission corridors, and ice storms that can collapse power lines require new approaches to system design, maintenance, and emergency response that can maintain service under increasingly challenging conditions.

Grid resilience investments are necessary to adapt electrical infrastructure to changing climate conditions while ensuring that critical services can be maintained during extreme events. These investments include underground power lines in high-risk areas, stronger poles and towers that can withstand severe weather, backup generation systems, and microgrids that can operate independently during major outages.

Heat stress on electrical equipment reduces its capacity and lifespan while potentially causing failures during peak demand periods when air conditioning loads are highest. Rising temperatures require derated operation of transformers, transmission lines, and generating equipment while potentially requiring costly equipment replacements or cooling systems that reduce overall system efficiency.

Water availability and quality issues affect power plant operations while creating competition for scarce water resources between electricity generation and other uses. Drought conditions can reduce hydroelectric generation while limiting cooling water availability for thermal power plants, requiring alternative generation sources or costly cooling system modifications.

Cybersecurity and Physical Security Threats

The increasing digitization of the electrical grid has created unprecedented cybersecurity vulnerabilities while traditional physical security threats continue to evolve, requiring comprehensive security strategies that address both digital and physical risks to critical infrastructure. The interconnected nature of modern electrical systems means that successful attacks on any component can potentially affect the entire grid while the consequences of major outages can include economic disruption, public safety risks, and national security implications.

Cybersecurity threats to the electrical grid have intensified as state actors, criminal organizations, and other malicious entities recognize the potential impact of successful attacks on critical infrastructure. The 2015 Ukraine power grid attack demonstrated that sophisticated adversaries can cause widespread blackouts through cyber means while more recent incidents have shown that ransomware and other cyber threats continue to evolve and proliferate throughout the energy sector.

The attack surface of the electrical grid is expanding rapidly as smart grid technologies, distributed energy resources, and Internet of Things devices create millions of potential entry points for cyber attacks. Many of these devices have limited security capabilities while being connected to networks that can provide access to critical control systems, creating vulnerabilities that are difficult to monitor and protect.

Supply chain security has become a critical concern as electrical equipment increasingly incorporates components and software from global suppliers that may not have adequate security controls or may be subject to foreign government influence. The potential for malicious code or hardware modifications in critical grid equipment creates risks that are difficult to detect and mitigate while the global nature of supply chains makes comprehensive security verification challenging.

Critical infrastructure protection requires coordination between federal agencies, state and local governments, and private industry to identify vulnerabilities, share threat intelligence, and coordinate responses to attacks or threats. This coordination is complicated by the complex jurisdictional authorities in the electrical industry while the private ownership of most grid infrastructure limits direct government control over security measures.

Physical security threats to electrical infrastructure include both terrorist attacks and vandalism that can cause significant outages while being relatively simple to execute. Transmission substations and power plants are particularly vulnerable targets that are difficult to protect comprehensively while the interconnected nature of the grid means that attacks on key facilities can affect much broader areas.

Insider threats represent another significant risk as employees and contractors with authorized access to critical systems and facilities may pose security risks through malicious actions, negligence, or compromise by external actors. Utilities must balance the need for employee access to perform their duties with security requirements that limit the potential for insider threats.

Technological Disruption and Innovation Opportunities

The electrical industry is experiencing rapid technological change that creates both opportunities for improved performance and challenges for traditional business models and regulatory frameworks. Emerging technologies including artificial intelligence, advanced materials, quantum computing, and biotechnology promise to transform how electricity is generated, stored, transmitted, and consumed while potentially disrupting established industry structures and relationships.

Artificial intelligence and machine learning applications are enabling new levels of optimization and automation throughout the electrical system, from predictive maintenance of grid equipment to real-time optimization of generation dispatch and demand response. These technologies can process vast amounts of data from smart grid sensors and systems to identify patterns and optimize operations in ways that were previously impossible while enabling autonomous responses to changing conditions.

Energy storage technologies continue to improve in performance while declining in cost, making storage viable for an increasing range of applications from grid-scale services to residential backup power. Advanced battery chemistries, compressed air energy storage, pumped hydro storage, and other technologies are expanding storage options while extending duration capabilities that could eventually enable seasonal energy storage.

Distributed energy resources including rooftop solar, small wind turbines, fuel cells, and energy storage systems are enabling customers to become active participants in the electrical system rather than passive consumers. These resources can provide system benefits by reducing peak demand and improving local reliability while creating new challenges for grid operations and utility business models.

Electric vehicle adoption is creating new demands on the electrical system while potentia lly providing new resources through vehicle-to-grid technologies that could use parked vehicles as distributed storage and grid support resources. Managing the grid impacts of widespread electric vehicle adoption while capitalizing on the opportunities they create requires new technologies, business models, and regulatory frameworks.

Advanced materials including superconductors, graphene, and nanotechnology applications could eventually enable revolutionary improvements in electrical equipment performance, efficiency, and capabilities. High-temperature superconductors could eliminate transmission losses while advanced conductor materials could increase transmission line capacity without requiring new right-of-way.

Quantum computing may eventually enable optimization capabilities that could revolutionize power system planning and operations while quantum communication technologies could provide unprecedented levels of cybersecurity for critical grid communications. While these applications remain largely experimental, they represent potential transformative technologies that could reshape the industry.

Regulatory and Policy Evolution

The rapid pace of technological change and evolving policy objectives require fundamental changes to regulatory frameworks that were designed for a different era of electrical industry structure and technology. Regulators must balance multiple objectives including reliability, affordability, environmental protection, and consumer choice while adapting to new business models and technologies that challenge traditional approaches to utility regulation.

Performance-based regulation is gaining acceptance as an alternative to traditional cost-of-service regulation that can better align utility incentives with customer and societal interests. These approaches tie utility returns to achievement of specific performance metrics related to reliability, customer satisfaction, environmental performance, or innovation while providing utilities with flexibility to achieve these objectives efficiently.

Market design evolution continues as competitive electricity markets adapt to high levels of renewable energy, energy storage, and distributed resources that have very different characteristics than the conventional resources these markets were designed to accommodate. New market mechanisms for capacity, ancillary services, and transmission planning are being developed to ensure adequate investment and reliable operations in systems with high levels of variable renewable generation.

Environmental regulations are driving rapid changes throughout the electrical industry as federal and state policies mandate reductions in carbon emissions, air pollutants, and water usage while requiring increased use of renewable energy sources. These regulations create both compliance costs and opportunities for utilities while potentially affecting the economic viability of existing facilities and infrastructure.

Customer choice and empowerment are becoming important policy objectives as distributed energy resources, energy efficiency technologies, and smart grid capabilities enable customers to have greater control over their energy consumption and sources. Regulatory frameworks must evolve to accommodate these capabilities while ensuring that the benefits and costs of grid infrastructure are allocated fairly among all customers.

Regional coordination and planning are becoming increasingly important as renewable energy resources, transmission planning, and system reliability require coordination across multiple jurisdictions and utility service territories. New approaches to regional governance and cost allocation are needed to support efficient resource development and grid operations while respecting state and local authority over energy policy.

Economic and Financial Challenges

The electrical industry faces unprecedented financial challenges as it must simultaneously invest in infrastructure replacement, grid modernization, renewable energy integration, and cybersecurity improvements while maintaining affordable electricity service and providing adequate returns to investors. These competing demands require new approaches to utility finance, regulation, and investment planning that can balance multiple objectives while managing financial risks.

Capital requirements for grid modernization and decarbonization are enormous, with estimates suggesting that trillions of dollars in investments will be needed over the next two decades to achieve policy objectives for reliability, sustainability, and resilience. These investment needs far exceed historical industry spending levels while occurring during a period of slow or negative load growth in many regions.

Cost recovery challenges arise as traditional utility business models based on selling increasing quantities of electricity face declining sales growth due to energy efficiency improvements and distributed generation while costs for infrastructure replacement and environmental compliance continue to increase. New rate designs and revenue models may be needed to ensure adequate cost recovery while maintaining customer affordability.

Stranded asset risks are significant as environmental regulations and market forces accelerate the retirement of conventional power plants and other facilities that may not reach the end of their useful lives. Utilities and their customers must bear the costs of these premature retirements while investing in replacement resources that can maintain system reliability.

Access to capital markets is essential for utilities that require substantial ongoing investments in infrastructure and technology, but utilities must maintain strong credit ratings and investor confidence while managing the risks associated with technological change, environmental regulations, and evolving customer expectations. Regulatory policies that affect utility returns and cost recovery mechanisms directly impact utilities' ability to attract the capital needed for necessary investments.

Competitive pressures from new technologies and business models can affect utility revenues and market positions in ways that create both opportunities and challenges. Distributed energy resources, energy service companies, and technology providers are creating new competition for traditional utility services while potentially creating opportunities for utilities to develop new revenue sources.

Opportunities for Innovation and Leadership

Despite the significant challenges facing the electrical industry, unprecedented opportunities exist for innovation, leadership, and value creation that could position the United States as a global leader in clean energy technologies and grid modernization. These opportunities require vision, investment, and coordination among industry stakeholders while potentially creating new industries, jobs, and economic benefits that extend far beyond the electrical sector.

Technology leadership opportunities exist in areas including energy storage, smart grid technologies, renewable energy systems, and advanced materials that could create export industries while improving domestic energy security and environmental performance. Government research and development programs, private investment, and university partnerships could accelerate technology development while creating intellectual property and manufacturing capabilities.

Grid modernization presents opportunities to create more efficient, reliable, and flexible electrical systems that can better serve customers while accommodating new technologies and business models. Successful grid modernization programs could serve as models for other countries while creating opportunities for American companies to export technologies and expertise.

Clean energy deployment opportunities include not only domestic renewable energy development but also the creation of manufacturing capabilities for wind turbines, solar panels, energy storage systems, and other clean energy technologies that could serve both domestic and international markets. These opportunities could create high-quality jobs while reducing dependence on foreign suppliers.

Workforce development opportunities exist as the electrical industry transformation requires new skills and capabilities throughout the workforce from engineers and technicians to managers and policymakers. Educational programs, training initiatives, and career development opportunities could ensure that American workers are prepared for the jobs of the future energy economy.

International leadership opportunities exist for the United States to influence global energy policy, technology standards, and climate action through successful domestic policies and technologies that can be exported and adapted to other countries' needs. American leadership in grid modernization and clean energy deployment could enhance international competitiveness while supporting global climate objectives.

The electrical grid transformation also creates opportunities for enhanced energy security, economic development, and quality of life improvements that extend far beyond the electrical sector itself. A modern, clean, and resilient electrical system can support economic growth, public safety, and environmental protection while providing the foundation for continued technological advancement and social progress.

The challenges facing the American electrical grid are significant and complex, but they also represent opportunities for innovation, leadership, and progress that could benefit society for generations to come. Success will require vision, investment, coordination, and commitment from all stakeholders, but the potential benefits justify the effort required to build an electrical system worthy of America's needs and aspirations.

Chapter 24: Conclusion - Powering the Future

The history of the United States electrical grid represents one of humanity's greatest technological achievements, transforming American society from an agricultural economy illuminated by candles and kerosene lamps into a modern technological civilization that depends on electricity for virtually every aspect of daily life. Over more than 150 years, the electrical grid has evolved from Thomas Edison's first power station serving a few dozen customers in lower Manhattan to a vast interconnected network that delivers electricity to over 300 million Americans while enabling the digital economy, modern healthcare, industrial production, and the quality of life that defines contemporary American society.

The Journey from Innovation to Ubiquity

The story we have traced through these chapters reveals a consistent pattern of innovation, adaptation, and transformation that has enabled the electrical industry to meet ever-changing societal needs while overcoming seemingly insurmountable technical, economic, and political challenges. From the early battles between alternating and direct current to the current challenges of renewable energy integration and cybersecurity, the electrical industry has demonstrated remarkable resilience and adaptability while maintaining its fundamental commitment to reliable, affordable service.

The War of the Currents that pitted Thomas Edison against George Westinghouse established alternating current as the dominant technology for long-distance power transmission, enabling the development of large-scale power systems that could serve entire regions rather than individual neighborhoods. This technical decision, made over a century ago, continues to shape the electrical grid today while providing the foundation for the interconnected systems that enable efficient resource sharing across vast geographic areas.

The rise of utility companies in the early twentieth century created the regulated monopoly structure that brought electricity to American cities and towns while establishing the principle of universal service that ensures electricity availability to all customers regardless of their location or economic status. The consolidation and standardization that occurred during this period created the economies of scale necessary for widespread electrification while establishing regulatory frameworks that continue to influence industry structure today.

Rural electrification programs of the 1930s and 1940s demonstrated the power of coordinated public policy to extend modern infrastructure to previously underserved populations, bringing electricity to millions of American farms and rural communities that private industry had considered uneconomical to serve. The success of these programs established electricity as a public necessity rather than a luxury while proving that government intervention could achieve social objectives that markets alone might not accomplish.

The post-World War II expansion period saw unprecedented growth in electricity consumption and generating capacity as the American economy boomed and electricity became integral to industrial production, residential comfort, and commercial activity. The development of large power plants, transmission networks, and distribution systems during this period created the foundation of the modern electrical grid while establishing electricity as an essential input for economic growth and social progress.

Lessons from Crisis and Change

The electrical industry's response to various crises throughout its history provides valuable insights into the sector's adaptability and the importance of learning from both successes and failures. The environmental awakening of the 1970s forced the industry to confront the air and water pollution caused by power generation while establishing environmental protection as a fundamental responsibility alongside traditional obligations for reliable and affordable service.

The energy crises of the 1970s demonstrated the risks of depending too heavily on imported fuels while highlighting the importance of energy efficiency, conservation, and diversified resource portfolios. The industry's response to these crises included investments in coal and nuclear generation that provided energy security benefits while creating environmental and safety challenges that continue to influence policy debates today.

The deregulation experiments of the 1990s and 2000s provided valuable lessons about the benefits and limitations of competitive markets in the electrical sector. While competition proved successful in some regions and applications, failures in California and other markets demonstrated that electricity's unique characteristics as a commodity require careful market design and continued oversight to ensure reliable service and consumer protection.

The Enron scandal and California energy crisis revealed the potential for market manipulation and abuse in deregulated electricity markets while highlighting the importance of transparency, oversight, and appropriate regulation in maintaining public trust and system reliability. The industry's response to these crises included enhanced market monitoring, improved governance structures, and more sophisticated regulatory frameworks that better balance competition with consumer protection.

More recent challenges including grid cybersecurity threats, climate change impacts, and renewable energy integration demonstrate the industry's continued need to adapt to changing conditions while maintaining its fundamental service obligations. The ongoing transformation toward clean energy and grid modernization represents the latest chapter in this story of continuous adaptation and improvement.

The Current Transformation

Today's electrical industry faces a transformation that is arguably as significant as any in its history, driven by the convergence of environmental imperatives, technological capabilities, and changing customer expectations that require fundamental changes to how electricity is generated, delivered, and consumed. The rapid growth of renewable energy sources is reshaping the generation mix while creating new challenges for system reliability and grid operations that require innovative solutions and unprecedented levels of coordination.

Smart grid technologies are enabling new levels of system monitoring, control, and optimization while creating opportunities for customers to become active participants in grid operations rather than passive consumers of electricity. These technologies promise to improve system efficiency, reliability, and customer service while enabling the integration of distributed energy

243

resources and new services that were not possible with traditional grid infrastructure.

The electrification of transportation, heating, and industrial processes is creating new sources of electricity demand while potentially providing new resources through vehicle-to-grid technologies and flexible load management programs. Managing these changes while maintaining system reliability and customer affordability requires new approaches to planning, investment, and operations that can accommodate rapidly changing conditions.

Energy storage technologies are enabling new applications and services while providing solutions to the intermittency challenges associated with renewable energy sources. The declining costs and improving performance of battery storage systems are making energy storage economically viable for an increasing range of applications while potentially transforming how electricity systems balance supply and demand.

The integration of information technology and artificial intelligence throughout the electrical system is enabling new levels of optimization and automation while creating cybersecurity vulnerabilities that must be carefully managed to maintain system security and public confidence. The digital transformation of the electrical grid represents both one of its greatest opportunities and one of its most significant risks.

Challenges and Opportunities Ahead

Looking toward the future, the electrical industry faces challenges that are both similar to and different from those it has confronted throughout its history. The fundamental challenge of providing reliable, affordable electricity service to all customers remains unchanged, but the context in which this challenge must be met has evolved significantly as society's expectations and dependencies have grown while new technologies and policy objectives create additional complexity.

Climate change and environmental protection requirements are driving unprecedented changes to the electricity generation mix while creating new demands for system flexibility and resilience that can accommodate variable renewable energy sources and extreme weather events. Meeting these challenges will require massive investments in new technologies and infrastructure while managing the transition away from existing facilities and systems.

Infrastructure aging and replacement needs create enormous capital requirements while presenting opportunities to deploy new technologies and improve system performance through strategic modernization investments. The decisions made about infrastructure replacement in the coming years will shape the electrical system for decades while affecting its ability to meet evolving customer needs and policy objectives.

Cybersecurity threats continue to evolve as the electrical grid becomes more digital and interconnected, requiring ongoing vigilance and investment in security measures that can protect critical infrastructure from increasingly sophisticated attacks. The challenge of maintaining security while enabling the connectivity and data sharing needed for modern grid operations requires careful balance and continuous adaptation.

Economic and regulatory challenges arise from the need to recover substantial infrastructure investments while maintaining customer affordability and providing adequate returns to investors. New business models, regulatory frameworks, and financing mechanisms may be needed to support the investments required for grid modernization while ensuring that the benefits and costs are allocated fairly among all stakeholders.

The Promise of the Future Grid

Despite these challenges, the opportunities for creating a better electrical system have never been greater. Advances in renewable energy, energy storage, smart grid technologies, and artificial intelligence are enabling capabilities that previous generations could only imagine while potentially creating an electrical system that is cleaner, more efficient, more reliable, and more responsive to customer needs than ever before.

A modernized electrical grid could provide the foundation for continued economic growth and technological advancement while supporting environmental protection and climate action goals that benefit both current and future generations. The electrification of transportation and other sectors could reduce air pollution and greenhouse gas emissions while creating new economic opportunities and improving public health.

Enhanced grid flexibility and resilience could reduce the frequency and duration of power outages while enabling faster recovery from major disruptions caused by natural disasters, cyber attacks, or equipment failures. Improved system monitoring and control capabilities could optimize operations in real-time while predicting and preventing problems before they affect customers.

Customer empowerment through distributed energy resources, demand response programs, and advanced rate options could provide individuals and businesses with greater control over their electricity costs while contributing to overall system efficiency and reliability. The democratization of energy production and consumption could create new opportunities for innovation and entrepreneurship while reducing the centralized control that has historically characterized the electrical industry.

Continuing Innovation and Adaptation

The history of the electrical grid demonstrates that continuous innovation and adaptation have been essential for meeting changing societal needs while overcoming new challenges and taking advantage of emerging opportunities. This pattern is likely to continue as the industry faces unprecedented change while building on the foundation of experience and expertise developed over more than a century of service.

Emerging technologies including advanced materials, quantum computing, biotechnology, and space-based solar power could eventually enable capabilities that seem impossible today while solving problems that currently appear intractable. The industry's track record of adopting and adapting new technologies suggests that these advances will be integrated into the electrical system as they become technically and economically viable.

International cooperation and knowledge sharing could accelerate the development and deployment of advanced grid technologies while enabling countries to learn from each other's experiences and avoid repeating costly mistakes. The global nature of climate change and environmental challenges creates incentives for collaboration that could benefit all participants while advancing common objectives.

Workforce development and education will be essential for ensuring that the electrical industry has the skilled professionals needed to design, build, operate, and maintain increasingly complex electrical systems. The transformation of the industry creates opportunities for new careers while requiring existing workers to develop new skills and capabilities throughout their professional lives.

The Enduring Importance of Electricity

As we conclude this historical survey, it is important to recognize that electricity has become so fundamental to modern life that its importance is often taken for granted until service is interrupted by storms, equipment failures, or other disruptions. The electrical grid represents one of the most complex and sophisticated technological systems ever created, requiring continuous coordination among thousands of participants while operating according to the laws of physics that allow no margin for error.

The men and women who have built and operated the electrical grid over the past 150 years deserve recognition for creating and maintaining this essential infrastructure while adapting to changing conditions and requirements. From the inventors and entrepreneurs who created the first electrical systems to the engineers, technicians, and managers who keep the lights on today, the electrical industry represents a remarkable achievement in human cooperation and technological advancement.

The electrical grid has enabled transformations in agriculture, manufacturing, communications, transportation, healthcare, education, and virtually every other aspect of human activity while creating opportunities for economic development, social progress, and improved quality of life that would not have been possible otherwise. The industries, jobs, and innovations that depend on reliable electricity service contribute trillions of dollars to the American economy while supporting the lifestyle and opportunities that Americans enjoy today.

A Living Legacy

The electrical grid is not a museum piece or historical artifact but a living, evolving system that continues to grow and change in response to new needs and opportunities. The history we have traced in these chapters provides context and perspective for understanding current challenges and opportunities while honoring the achievements of those who created the foundation on which today's electrical system is built.

The decisions being made today about grid modernization, renewable energy integration, and regulatory policies will determine how well the electrical system serves future generations while building on the legacy of innovation, service, and adaptation that has characterized the industry throughout its history. The responsibility for continuing this legacy falls to current and future leaders in government, industry, and academia who must balance multiple objectives while maintaining the reliability and affordability that society has come to expect.

The story of the American electrical grid is ultimately a story about human ingenuity, perseverance, and cooperation in creating and maintaining essential infrastructure that serves the common good. It demonstrates what can be accomplished when technical expertise, economic resources, and political will are mobilized to address important societal needs while adapting to changing conditions and emerging challenges.

As the electrical industry continues to evolve in response to environmental imperatives, technological capabilities, and customer expectations, the fundamental principles that have guided its development—reliability, affordability, universal service, and continuous improvement—remain as relevant today as they were when Thomas Edison first illuminated Pearl Street. The future grid will look very different from the system we have today, but it will continue to serve the same essential function of powering American society while enabling continued progress and prosperity.

The history of the United States electrical grid is not yet complete. The most important chapters may still lie ahead as the industry faces the challenges and opportunities of the 21st century while building on the solid foundation created by previous generations. The success of this ongoing transformation will depend on the wisdom, dedication, and vision of those who carry forward the responsibility for this essential infrastructure while ensuring that it continues to serve all Americans reliably, affordably, and sustainably for generations to come.

In the end, the electrical grid represents more than just wires, transformers, and power plants—it represents the collective commitment of American society to providing essential services that enable opportunity, prosperity, and progress for all. This commitment has driven the remarkable achievements chronicled in this history while providing the foundation for continued success in the challenges that lie ahead. The story c

www.ingramcontent.com/pod-product-compliance
Lightning Source LLC
Chambersburg PA
CBHW070554130626
46556CB00001B/153